Earth Science Resources in the Electronic Age

Earth Science Resources in the Electronic Age

Judith A. Bazler

Science Resources in the Electronic Age

Greenwood Press
Westport, Connecticut • London

Library of Congress Cataloging-in-Publication Data

Bazler, Judith.
    Earth science resources in the electronic age / Judith A. Bazler.
       p. cm.—(Science resources in the electronic age)
    Includes bibliographical references and index.
    ISBN 1–57356–381–1 (alk. paper)
    1. Earth sciences—Computer network resources. I. Title. II. Series.
QE48.87.B39   2003
025.06′55—dc21   2003052843

British Library Cataloguing in Publication Data is available.

Library of Congress Catalog Card Number: 2003052843
ISBN: 1–57356–381–1

First published in 2003

Greenwood Press, 88 Post Road West, Westport, CT 06881
An imprint of Greenwood Publishing Group, Inc.
www.greenwood.com

Printed in the United States of America

The paper used in this book complies with the
Permanent Paper Standard issued by the National
Information Standards Organization (Z39.48–1984).

10 9 8 7 6 5 4 3 2 1

CONTENTS

ACKNOWLEDGMENTS

The author extends her appreciation to the graduate students at Monmouth University in West Long Branch, New Jersey. I acknowledge their assistance in researching and reviewing Earth Science Web sites, their support and enthusiastic interest in this manual and in the larger process of infusing technology into their instruction and into their everyday lives. A number of students reviewed the first chapter for relevance and usability. In addition three graduate assistants formalized the review process and provided a number of the Web sites that the author reviewed for addition in the manual. Specifically I wish to thank the following graduate students of Monmouth University: Cynthia Coughlin, Julia D'Alessandro, Dena DeFlora, Erica Golterman, Danielle Graham, Ryan Hamilton, Anne Hazeldine, Michael Iasparro, Colleen Kenny, Michael Lozinski, Karen Magaraci, Allison Meyer, Bradley Millaway, Jennifer Orgo, Elizabeth Rogers, Laurie Rosenthal, Lisa Ruggiero, Laura Ryan, Christine Tortoriello, Audra Trost, Lynn West, Cheryl Whinna, Mark Alfone for reviewing Chapter 1, and graduate assistants Hasmukh A. Patel, Abdelghani Lakmini, and Jeff Portland.

My respect and admiration goes to Dr. Eleanora Von Dehsen, whose vision and leadership led to the development of this manual and the series that follows. Finally, I dedicate this book to my mother and father, who encouraged me to "get an education," and to my son, Kirk Melnikoff, who has always been my greatest friend.

INTRODUCTION

Earth Science Resources in the Electronic Age is designed as a one-stop source for cutting through the chaos of the Internet to find authoritative information on topics covered in the Earth Science curriculum. The book is divided into five chapters. In the first, "The Basics," you'll find an invaluable introduction to the different kinds of electronic media. You might be surprised to learn that there's more than just the Web on the Internet. In addition to learning about the types of electronic resources, you'll also see how specific kinds of Web sites can be mined for your research projects. This chapter also provides basic information about the Web, search engines and other search tools, evaluating Web material, and copyright and plagiarism issues that are unique to today's electronic resources.

The heart of the book, Chapter 2, "Resources in Earth Science," provides you with a treasure map to high-quality information on the Web, which will save you hours of your own research time. Here, we first point you to the Web's top-notch sites offering general information about science. Next, we have searched for and found the crème de la crème of Web sites providing specific information on key topics in Earth Science. These topics are listed alphabetically rather than by the name of the Web site so that you can immediately go to the information you need without guessing if a site will be useful.

For each topic, you'll find reviews of several Web sites, giving you all of the information you need to know: the name, URL, appropriate grade range, and a thorough discussion of how to use the site for research. When you log on to the Web to find background information on fossils or to gain a fuller understanding of erosion, you'll now have a number of handpicked sites, as opposed to the thousands that might turn up with a keyword search. In case you choose to conduct your own online search for a key topic, we let you know which search engine and keywords provide the best "hits."

Chapter 3, "Supplies," reviews a number of excellent Web sites that offer materials and resources for parents and for science students, including the premier science supply companies, such as Flinn Scientific, Sargeant-Welch, and others.

Chapter 4, "Museums, Science Centers, and Summer Programs," surveys Web sites offering unique online museum exhibits, interpretive centers, summer programs, and other interactive opportunities for students of science.

The final chapter of the book, "Careers," turns its attention to Web sites that provide students with career information in the field of science. Here, we've reviewed sites for professional associations, academic groups, conferences, workshops, programs, clubs, and other outlets for students interested in working or doing an internship in the subject.

SCOPE AND CONTENT

We gathered the topics listed in Chapter 2 from a detailed analysis of the science standards and leading Earth Science texts. The index will lead you to topics not on the list in Chapter 2 (see below).

The number of Web sites reviewed in each entry is determined by the number of good Web sites that look at different aspects of the topic, not the importance of the topic. We reviewed as many of the Web sites on each topic as possible and chose to review as many as possible. Most topics provided a number of excellent Web sites that are included in this book. Obviously, some topics had many Web sites to review and only a few are included in this book due to space limitations. We did not limit the number of sites per topic, and we have chosen the Web sites based on how closely they mirror what is covered in textbooks and the national science curriculum.

HOW TO USE THIS BOOK

There are three ways you can find information in *Earth Science Resources in the Electronic Age*. First you can look at the detailed table of contents, which will quickly lead you to such topics as "copyright and plagiarism" or "existing program sites." If you are researching a particular topic in Earth Science, you can immediately go to the alphabetical listing of topics in Chapter 2. Finally, you can use the index, which expands our coverage significantly. Because we had to limit the number of topics in Chapter 2, we added as much detail as possible to our site reviews. These include other topics covered in the Web site but not included in our topic list. All of these have been indexed. For example, "cumulus" is not in the topic list, but if you look in the index, you will find that a site about clouds discusses them. Therefore, if you don't find what you want in the topic list, go to the index.

So, when you are desperate for a quick, reliable Web site for a report on plate tectonics; need to know about Internet searching or citing electronic sources; or want to know what it's like to be a geologist, reach for *Earth Science Resources in the Electronic Age*. It will help you avoid the frustration of endless surfing and wasted time by taking you directly to the Web's best Earth Science resources.

1
THE BASICS

INFORMATION IN THE ELECTRONIC AGE

Today, you are just as likely to turn to a computer terminal as to a book to answer your science reference questions. In fact, the Internet is considered by most to be the "preferred" reference tool. Unfortunately many budding researchers use a hit-or-miss approach to their research strategy. This approach uses a lot of time and even more patience. If you love the challenge, this "discovery" approach may be exactly what you need to feel like an explorer. However, for the majority of researchers who find the hit-or-miss approach time consuming, this lack of a systematic process may lead to a high degree of frustration. The approach you choose to use is not the only barrier you must overcome in order to feel comfortable and confident in your searching. Even if you are successful using either a hit-or-miss search process or a more systematic process, the multitude of information gathered and the task of eliminating useless material without any standardization or guide may make the task of researching a topic on the Internet seem overwhelming and unachievable. It is at this point when more is not necessarily better. You may wonder how it is that we evolved from using only library card-catalog indexes and typewriters to using electronic computers and the Internet in our searching strategies.

The Internet initially provided information exchange between scientists. In the beginning, the Internet was owned and managed by the U.S. government. This governmental ownership changed in 1995 to a predominantly privately controlled organization. The Internet originally "internetted" university-based research centers with governmental contractors, specifically military and defense centers. This initial network, developed in 1969, was called ARPANet (Advanced Research Projects

Agency Network, of the Department of Defense). Early researchers concerned themselves with communication and information exchanges and resources. They were able to use different computers as long as each computer followed a set of rules for expressing the information. These rules are called "protocols," and Telnet and Gopher are a few examples of early protocols of the electronic age. In these earlier protocols, graphics and illustrations were not easily transmitted. The Internet now commonly uses TCP/IP (Transmission Control Protocol/Internet Protocol). Where does the World Wide Web (WWW) fit into this picture? Is the Internet the same as the World Wide Web? The Internet provides tools for communication, inquiry, and construction. It is much larger than just the World Wide Web and includes newsgroups, chat rooms, mailing lists, videoconferencing and e-mail, to name a few uses.

The Internet's versatility, combined with multimedia capabilities such as sound and graphics, makes it one of your research tools of choice. You would think that a tool that is so heavily used by so many people would have minimal problems or deficiencies. Even though a wealth of information can be found on the Internet, the lack of evaluative standards, the confusion of researching a topic, inactive or "dead" Web sites, and the ever-changing product upgrades will plague you when searching the Internet. Yes, using electronic resources can be extremely exasperating to everyone.

Remember, currently the Internet provides access to only a small percentage (10–15%) of all usable reference material. Since the Internet can provide access to only a small amount of all reference material available, you will need to know about other formats of science reference material, the availability and the value of these references. A science researcher uses all formats of reference material when fully exploring a science topic. How and where do you find answers to questions on specific topics taught in science courses? The *Science Resources in the Electronic Age* series should be the first reference before you begin your in-depth research of any science subject or topic.

Earth Science Resources in the Electronic Age is the third volume of an open-ended series that helps researchers quickly find reliable, age- or level-appropriate information on biology topics using the Internet. It is designed for students from junior high through the first years of college as well as the general reference audience and it offers you an expert guide to the vast electronic information resources available. The series combines searching strategies and annotated sites to help you use electronic resources effectively. Because Internet information is often incomplete, the series also shows how other electronic media and print materials can supplement the Web.

In addition to *Earth Science Resources in the Electronic Age*, other volumes include *Chemistry Resources in the Electronic Age*, and *Biology Resources in the Electronic Age*.

Each volume in the series is divided into five distinct sections. The first chapter, "The Basics," contains a general overview of research in the electronic age and an introduction to electronic technologies. The chapter also includes information on different formats, Web basics and search strategies, and evaluation tools. The focus is on research in the particular discipline covered in the volume. For example, the focus of *Earth Science Resources in the Electronic Age* is Earth Science.

The second section, "Resources in Earth Science," is the heart of the book. It presents the resources available on key topics in the discipline. Each category begins with a quick-search feature, listing the most appropriate metasearch engine for researching the concept and suggesting the most useful keywords to find the appropriate questions generated by the science curriculum and for their scientific accuracy. Emphasis is on the Internet, although other electronic media and print materials will be discussed. Each source citation includes the name and URL of the site, its appropriate grade or age range, and a brief review of the site.

The remaining sections of each volume expand the scope of the book, offering information specifically for educators, resources for special-needs children, places from which to order scientific supplies, a complete list of museum sites, and information on careers.

FORMATS OF RESOURCES

Science resources can be obtained in many formats, including the Internet. In other words, science help can be found in printed forms such as books, journals, and curriculum, on CD-ROMs and videos, and in various professional organizations, science museums, and private corporations. The following discusses these formats and provides general resource addresses for further information.

Library Electronic Services

A number of academic and professional directories offer a collection of Internet resources. In addition, specific commercial portals, such as gonetwork (http://www.go.com) and open directory project (http://

www.dmoz.org) provide a reference directory that can be searched using author, subject, or title. To obtain a list of libraries that have Internet-accessible catalogs use LIBWEB at http://sunsite.berkeley.edu/Libweb. Another usable portal is the Library of Congress, accessible at http://www.loc.gov/library.

CD-ROM/ Multimedia

A limited number of CD-ROMS, videos, and cassette tapes are specifically concerned with science and more specifically biology. In addition, libraries use CD-ROMs in order to access large collections. Some CD-ROM indexes used by libraries are *Reader's Guide to Periodical Literature*, *Books in Print*, *InfoTrac*, *ProQuest's Periodical Abstracts*, *Academic Search* or *Alternative Press Index*. You can also purchase CD-ROM reference materials, which contain videos, animations, pictures, and sound clips. *World Book Encyclopedia* is an example of a multimedia CD-ROM.

E-Mail

E-mail, which provides you with a quick method to exchange information with others, was the earliest function of the Internet. In order to use the Internet as an e-mail message tool, you simply need to have an account through an e-mail service provider. Once connected, you need to know the e-mail address of the person(s) you are trying to send a message. Yahoo! People Search at http://people.yahoo.com or at http://ussearch.com can help you search for the address of a person. Various sites provide "expert" help and advice to you about science subjects. The Center for Improved Engineering and Science Education (CIESE) provides links to ask-an-expert sites at http://njnie.dl.stevens-tech.edu/askanexpert.html.

Mailing Lists

Mailing lists are similar to e-mail; however, the contact in a mailing list is not a single individual but a group of individuals with similar interests. In order to participate in a mailing list you must subscribe to one. Listserv@msu.edu is a mailing list on Educational Technology, listserv@unm.edu is a mailing list on integrating technology in schools, and listserv@postoffice.cso.uiuc.edu is a list on middle-school topics. For a complete list of mailing lists use Liszt at http://www.topica.com. Once at the site, type in "science" at the search window. This leads you to a huge list of national and regional mailing lists in science.

Usenet Newsgroups

A newsgroup maintains its messages on a bulletin board like server. Messages can be posted to the bulletin board by anyone who has registered. In order to obtain a directory of Usenet newsgroups go to http://www.deja.com/usenet or http://www.cyberfiber.com/index.html and scroll to the "science" link. After accessing the science page, you will be given a list of groups with an indicator of the amount of activity occurring with the groups. You can read the messages, but in order to post a message you must simply register.

Chat

Chat rooms provide you with "real-time" ability to "talk" directly to people. Chat rooms differ in the ability to be synchronous, real-time, but newsgroups and e-mail are not in real-time. Chat rooms are like face-to-face or telephone discussions. Science Live Chat from Western Canon University at http://mobydicks.com/commons/Sciencehall/live/chat.cgi is a live chat for one hour every day where people discuss great books of science. This classroom is active from 9:00 P.M. to 3:00 A.M. Eastern time. Other chat rooms can be found by searching using "science chat rooms" for the searching phrase.

Videoconferencing

Videoconferencing allows participants to see and hear each other. Cornell University originally designed the CU-SeeMe software. NASA uses this format to provide live feeds during space shuttle missions. The Global SchoolNet Foundation maintains a list of CU-SeeMe schools at http://www.gsn.org/cu/index.html.

Full-Text Magazines and Newspapers

A number of resources are available to the public that can be purchased or used through the library. Newspapers have become a wonderful resource for current science development. The *New York Times'* Science Times section is my favorite resource for up-to-date science news. The *Los*

Angeles Times also has an excellent science section. *Television News* at http://www.televisionnewscenter.org and *CNN Science and Technology* at http://www.cnn.com are also excellent resources for science research. Magazines (a list can be found at Enews: the ultimate magazine site, which is at http://home1.gte.net) including *Discover, Popular Science, National Geographic World, Audubon,* and *Natural History* provide current science written in a usable, friendly manner. A more ambitious researcher might look into the professional journals such as *Science* and *Scientific American.* A complete list of professional journals can be found in the Electronic Journal Index at http://www.e-journals.org.

Digital Libraries

Libraries are busy storing whole collections on computers. The American Memory Learning Page provided by the Library of Congress at http://lcweb2.loc.gov/ammem/ndlpedu opens the door to collections, activities, and lessons and resources for you mostly in social studies but also in inventions and technology. Single-subject libraries exist, such as the Banting Digital Library found at http://newtecumseth.library.on.ca/banting. Other sites of interest are:

http://collections.ic.gc.ca Canada's Digital Collections
http://moa.cit.cornell.edu/dienst-data/cdi-math-browse.
     html Cornell Digital Library Mathematics Collection

TYPES OF WORLD WIDE WEB RESOURCES AVAILABLE

The World Wide Web contains numerous science Web sites that will provide valuable information/tutorials, experiments/activities, science materials, and other reference materials. There are so many that you probably could use some help in choosing the best general sites in science. The following contains the names of reference materials in science, commercial and government resources, academic and educational resources and existing science programs. In addition, I have included a brief discussion about practical activities and demonstration resources.

Reference and Journals Web Sites

The Web contains a number of wonderful reference sites that contain not only resources that are found in dictionaries and encyclopedias but also links to other great resources. These sites are usually well organized and updated regularly.

Title: StudyWeb
URL: http://www.studyweb.com
Grade Level: Any grade level
Review: StudyWeb has officially become part of The Lightspan Network and is now available only by school subscription. This site lists more than 162,000 URLs. You search by selecting a topic, typing in a topic, or clicking on a specific topic list. If you click on the "biology" link you are led to classroom resources, history of biology, branches of study, teaching resources, and educational and professional development. I clicked on the branches of study topic of "polymers," which led me to a list of URL sites that are rated by apples (four is excellent) for visual content. The site has the approximate grade level, source, and contributor name. A brief review of this site is included.

Title: InfoPlease
URL: http://www.infoplease.com
Grade Level: Any grade level
Review: A wonderful reference tool. You can search the almanacs or the almanac index. The two that are directly appropriate to science are Health and Science and Weather and Climate. You also can type in a topic at the search window. This site has a search window for topics and one for biographies. I typed in the word "Einstein" at the biography search window and was linked to encyclopedia, dictionary, and almanac links. The almanac link provided a picture, birth/death dates and birthplace. The encyclopedic link provided information on his life, contributions to science, his writings, and a bibliography. This site also has a link to a pronunciation key!

Title: Britannica.com
URL: http://www.britannica.com
Grade Level: Any grade level
Review: A very easy site to navigate. You type in the topic at the beginning search window, alphabetically, by subject or by world atlas. I explored by typing in the terms "acid rain," which led me to Web site links that are rated (four stars is excellent and recommended), current event and magazine links, as well as a great encyclopedic reference complete with video/animation. There also is a [+] icon

throughout the site to enable you to provide feedback to the site's developer. At the main menu is a link to Britannica School, explaining that if you subscribe to this component of the site there are further curricular materials available. I found the free component to be a worthwhile resource.

Commercial and Government Web Sites

Do not avoid commercial sites because you think that these sites contain only science materials for sale. Carolina Biological, Pasco Scientific, Edmund Scientific, Educational Innovations, Inc., and Fisher Science Education On-line not only provide a source for you to buy a needed piece of equipment or that lens you must have but also offer practical information and activities. In addition, specific government programs from the National Science Foundation (NSF), and U.S. Department of Education (USDE) provide the researcher with grant possibilities, program information, and other science resources.

Academic and Educational Web Sites

Professional organizations are a wonderful resource for information, science activities, and science connections. The American Chemical Society (ACS), American Association of Physics Teachers (AAPT), National Science Teachers Association (NSTA), National Association of Biology Teachers (NABT), and National Council of Teachers of Mathematics (NCTM) are only a few of the academic/educational sites that are a "must" for the science researcher.

Many professional organizations maintain a national and/or international presence through group conferences and professional journals. Science professionals use both the journal and conference forum in order to share research and build collaborative partnerships. In science, the American Association for the Advancement of Science (AAAS) and the National Academy of Science (NAS) are considered to be the most prestigious science organizations. Both publish journals and information for those interested in science and as a forum for science professionals to share both information and research.

Title: The American Association for the Advancement of Science
URL: http://www.aaas.org
Grade Level: Any grade level

Review: The American Association for the Advancement of Science organization publishes the journal *Science* and provided the forum needed for the development of Project 2061. The AAAS Web site links you to:

- National/International science conference and meetings dates
- Science policy
- Science workshops and programs (national/international) including radio programs that focus on science topics
- An index of science resources including science books, videos, CD-ROMs, and science products
- The latest research in science
- Science shows, including *Kinetic City Super Crew*
- A grant searchable database
- Links to resources including Project 2061 books and CD-ROMs

Title: The National Academy of Science (NAS)
URL: http://www.nationalacademies.org
Grade Level: Any grade level
Review: The National Academy of Science (NAS) is one of the branches of the National Academies organization that provides information on advisers to the nation on science, engineering and medicine. The other organizations in the National Academies are the National Academy of Engineering, Institute of Medicine, and National Research Council. The NAS Web site links you to its publication, *Proceedings of the National Academy of Science*, and to the following:

- Special science programs
- Science education
- International science policy
- Meeting dates of science conferences
- The latest science research

In addition, a quick review of the National Research Council (NRC) link leads you to the National Academy Press, with access to over 1,350 books online, including a table of contents for you to use to make your selection and ordering information. The NRC link also gives you access to both the NRC and NAS archives, providing you with information and access to past research.

Overall, both the NAS and AAAS sites provide the researcher with the latest science research and development. The best way to access them is directly through the above URLs.

Title: National Science Teachers Association (NSTA)
URL: http://www.nsta.org
Grade Level: Any grade level
Review: Along with NAS and NRC, both the National Science Teachers Association (NSTA) and the National Association of Biology Teachers (NABT) provide educators with the latest developments in biology education. A search of the NSTA Web site leads you to its journals (*Science Scope, The Science Teacher, Journal of College Science Teaching, Quantum, Dragonfly,* and *Science and Children*) and the following:

- Position statements and policy on science education
- National student competitions
- Science clubs and organizations
- Science resources
- U.S. registry of science teachers
- Science workshops for educators
- Science awards
- Scilink, which is a free service that connects a Scilink textbook to the Web
- A science video vault
- Science of energy
- A discussion room about building a presence for science education

Title: National Association of Biology Teachers
URL: http://www.nabt.org
Grade Level: Any grade level
Review: The National Association of Biology Teachers (NABT) is "the leader in life science education." The NABT empowers educators to provide the best possible biology and life science education for all students by providing expertise and opportunities for members to enhance their professional performance, advocating the teaching and learning of the biological sciences based on the nature and methods of science and the best practices of education, attracting and representing the full spectrum of educators in biology and the life sciences, and operating with benchmark levels of organizational effectiveness and efficiency. The home page has links to News & Events, NABT

Store, Members Only, Conferences & Workshops, Publications, Resources, and Education & Professional Development. You can peruse the links or type in a topic in the search window. This is the premier Web site for biology teachers.

Existing Programs Web Sites

A number of existing science programs are available to the researcher. Programs can be found in the government resources if funded by the government. The Jason Project is just one of these federally funded programs. In addition, publicly funded programs, such as the Westinghouse Science Talent Search at http://www.westinghouse.com, are useful to the researcher.

Practical Activities Web Sites

My favorite resource to find practical activities is the Eisenhower National Clearinghouse for Mathematics and Science Education. Another useful tool is the Lesson Plan Search (ERIC) or Ask ERIC (Educational Resources), found at http://www.eric.ed.gov.

WEB BASICS

The World Wide Web (WWW) has enhanced the use of the Internet by using both multimedia and hypermedia. The multimedia capability allows you to communicate by integrating text, video, audio, and still images into a single product. The hypermedia capabilities of the WWW allow you to experience the multimedia in any order. The WWW uses its multimedia and hypermedia capability and displays these as pages and links on a computer screen. The page has text, images, graphics, and video clips and the links take you to other sources of information.

The Web uses Hypertext Transfer Protocol (HTTP) as its primary protocol. Other protocols are File Transfer Protocol (FTP), Network News Transfer Protocol (NNTP), Telnet, Gopher, and Simple Mail Transport Protocol (SMTP). The WWW uses a group of Internet servers. The Web browser is the program that allows you to converse in these different protocols.

Browsers

A browser provides the tool that you need to access the World Wide Web and other information resources. Consider the browser to be like a workshop in which you use different tools to do different jobs. The two most popular browsers are Microsoft Internet Explorer (Explorer) and Netscape Communicator, in which Netscape incorporated Netscape Navigator. Browsers provide different tools for doing different jobs while using the World Wide Web. For instance, most browsers have a Back button, which allows you to return to a page that you just visited. In addition, most browsers contain the following buttons:

- Back—allows you to go back to a page
- Home—returns you to your home page or to your browser's home page
- Reload or Refresh—reloads the Web page
- Search—takes you to a list of search engines or directories that are specific to the Web browser
- Bookmarks or Favorites—stores a URL for future reference
- Print—allows you to print a document

Under the tools, you will find a box labeled "Location," "Go To," or "Address." This is where you will type the specific Web address, called a URL.

URLs

You should think of URLs (Uniform Resource Locators) as addresses of Web sites that contain information. You probably have a home address, which usually includes your name, number of the building, street, city, state and zip code. URLs are very similar to home addresses. Some of you may also have an e-mail address, which looks like the following: user@somewhere.domain. A typical Web site address will look like the following: a computer's name@somewhere.domain. An example of a URL address, for a Web site for cyberspace law, is: http://www.cs.csubak.edu.

- HTTP://—hypertext transfer protocol, the beginning of a URL (in front of slashes), indicates either the resource or method of access. The most common research protocol for Web file transfers is http. Others include Gopher—The gopher search tool; FTP—a file available for downloading; News—newsgroups; Telnet—another computer system; WAIS—Wide Area Information Search database; and File—your own file.
- WWW—World Wide Web, host computer name
- cs.csubak—specific computer or server
- edu—educational (domain designator), the last part (after the dot) of the URL contains the specific address or DNS (domain name system). The last three letters after the period of a URL indicate the type of group that owns

the server. In our example, the last three letters are ".edu," which stands for "university." Other designations are: .com = business or commercial, .org = organization, .mil = military agency, .net = network provider, .gov = U.S. government, .au = Australia .us = United States, .ca = Canada, .fr = France, .uk = United Kingdom, .k12__.us = U.S. public schools. The Internet Corporation for Assigned Names and Numbers (ICANN) sets up domain names.

Research on the Web is performed through a process called a *search*.

Metasearch Engines

In order to find URL addresses that are specific to the topic that you are researching, you need to use the appropriate computer "tool." A searching tool that uses multiple search engines or directories is called a metasearch engine (also called multithreaded engines). Metasearch engines help speed up the process of a search. You can consider metasearch tools to be the Swiss Army Knives of search engines in that they provide many tools to do the job. Examples of metasearch engines are found in Table 1.1.

Metasearch engines are great to use when first researching a subject. You probably are now wondering which one or ones of the above to use in your search. Currently, high praise has been given to Metacrawler, MetaFind, and Dogpile from reviewers of metasearch engines. My university students tend to use AskJeeves because they are able to search with no specific searching rules, and once you do a search, the tool begins to ask you a bunch of questions to enable you to narrow your search. I personally use Metacrawler because of its ease of use, its removal of duplicate sites, and its ability to interpret Boolean search rules, and because it searches with the best search engines. After you have searched a topic/subject and you find that you need more detail or more information, you may want to use an individual search engine in order to further refine your search.

Search Engines

Search engines help you research various subjects or words. If the browser is your workshop, consider both metasearch engines and search engines to be your tools to obtain information. Generally, there are different search tools available depending upon the method that you choose to use in your search. There are directory search engines, which find infor-

Table 1.1
Metasearch Tools

Metasearch Name	URL	Description
All-in-One	http://www.allonesearch.com	
Ask Jeeves	http://www.askjeeves.com	A simple tool. Very limited.
Cyber411	http://www.cyber411.com	Uses 16 search engines. Uses Boolean logic for searching and eliminates duplicates.
Debriefing	http://www.debriefing.com	Searches Alta Vista, Infoseek, Excite, Webcrawler, Lycos, and HotBot. Uses Boolean logic, ranks the sites, and eliminates duplicates.
Dogpile	http://www.dogpile.com	Searches 14 search engines, newsgroups, business news, and newswires. Uses Boolean logic for searching.
Highway 61	http://www.highway61.com	Searches Yahoo!, Lycos, Webcrawler, Infoseek, and Excite. Uses Boolean logic and tends to be slow.
Inference Find	http://www.infind.com	Searches Yahoo!, Alta Vista, Webcrawler, Infoseek, and Excite. Eliminates duplicates.
Internet Sleuth	http://www.thebighub.com	You determine which search engines it uses.
Mamma	http://www.mamma.com	Searches the Web, Usenet, news, stock symbols, company names, MP3 files, pictures, and sound; also searches Alta Vista, Excite, Infoseek, Lycos, Webcrawler, and Yahoo!. Uses implied Boolean logic (+/−).
Metacrawler	http://www.metacrawler.com	Searches Lycos, Infoseek, Webcrawler, Excite, Alta Vista, and Yahoo!; also searches computer products, Usenet, files, and stock quotes. Uses implied Boolean logic (+/−) for searching.

Table 1.1
(*Continued*)

Metasearch Name	URL	Description
Metafind	http://www.metafind.com	Searches Alta Vista, Excite, HotBot, Infoseek, Planetsearch, and Webcrawler. Uses Boolean logic.
Profusion	http://www.profusion.com	Searches the Web or Usenet. Uses Boolean logic, tends to be slow.
Savvysearch	http://www.savvysearch.com	Uses many search engines. Uses Boolean logic, tends to be slow.
Verio Metasearch	http://search.verio.net	Uses many search engines. Tends to be slow.

mation by using subjects. There are also keyword search engines, which search by specific words. What is most confusing about the method of searching is that most will do both a search by keywords and a search by subject. In addition, you may find that your information specifies that there are subject directories and search services. Most search engines will have a menu of subjects to click on or a space in which you provide the subject or keyword. The problem should be which search engine to use and how to use it. There are more search engines on the World Wide Web than will be included in this book. Yahoo!, AltaVista, HotBot, Infoseek, LookSmart, Google, and Excite are some of the more common. Table 1.2 lists search engines and states their URL addresses, whether they use Boolean or limited logic, and the size of the search.

Other Tools

A great Web site that provides more current information on the Internet is Zen and the Art of the Internet Web site at http://www.cs.indi ana.edu/docproject/zen/zen-1.0_toc.html. If the Web site that you are researching contains video, images, or sound, it will direct you to a helper piece of software called a plug-in. The plug-in works with the Web browser, enabling you to use the multimedia required in the Web site. Files that require helper software are called MIME (Multimedia Internet Mail Extension) files. Adobe Acrobat Reader is a typical plug-in. It al-

Table 1.2
Search Engines

Web Site URL	Boolean Logic	Search Size
Google http://www.google.com	AND assumed No OR	Medium–large
Alta Vista http://www.altavista.com	AND, OR, AND NOT, NEAR	140–160 million
Northern Light http://www.northern.com	AND, OR, NOT	140–160 million
Infoseek http://infoseek.go.com	NO +/–	30 million+
FastSearch http://alltheweb.com	NO +/–	200 million+
Librarians' Index http://www.lii.org	AND implied, OR, NOT	5,000
Infomine http://infomine.ucr.edu	AND implied, OR	16,000
Britannica Web's Best http://www.britannica.com	AND, OR, NOT	150,000
Yahoo! http://www.yahoo.com	NO +/–	1 million
Galaxy http://galaxy.einet.net	AND, OR, NOT	300,000

lows you to see PDF (Portable Document Format) files. This plug-in is available for free at http://www.adobe.com/products/acrobat/readstep. html. Most plug-ins are available free. Real Audio 4.0, Real Video 4.0, QuickTime, (WAV, MP3, and AVI are extensions) are more examples of helper software. Real Player and Real Jukebox allow you to see action or hear sounds live on the Internet. These plug-ins are available at http://www.real.com. QuickTime (http://www.apple.com/) is the plug-in for use for movies and videos.

Where plug-ins have expanded your ability to use the Web, new programming languages such as Java, Java Applets, JavaScript, Virtual Reality Modeling Language (VDRML) and Extensible Markup Language (XML) have expanded the Web abilities. Java is a programming language developed by Sun Microsystems in order for you to work on any PC.

JavaScript, a program that works with hypertext markup language (HTML), is three intermingled languages.

You should try some of the current tricks that are becoming standard in the Internet world. Try typing the word Finger and a DNS. For example, finger@monmouth.edu or just Finger at the location window. This will get you information about users. You can also talk to someone without going into email by typing Talk (the person)@DNS; for example, Talk Bazler@monmouth.edu. If I were on the computer at the time we could proceed to talk without accessing e-mail. This last development will help you with a search if you know the name of the person or if you know the domain of the institution. You type WHOIS at the location window. For example, if you are looking for a person's name, type whois@mit.edu. This will provide you the names of registered users at the Massachusetts Institute of Technology. Or if you know the name but not the domain, type Whois Smith if a user is registered, and a Web site will be recognized.

These are just a few of the latest new tools available for searching in a different way, not using a search engine. However, the process of researching a topic isn't over once you've chosen the appropriate tool (metasearch/search engine). The following suggested process will give you the steps you need to communicate your topic to the search engine.

SEARCHING

One of the most difficult stages of a search occurs when you attempt to communicate your needs precisely to the search engine. The following is a suggested step-by-step process for refining your language for typing into the search window:

- Write a sentence(s) stating what you want to research.
- Identify keywords. Underline those in your sentence(s).
- List synonyms of keywords.
- Combine synonyms and keywords with or/and/not or +/–. Put parentheses around "or" phrases.
- Check spelling.

For example, If you want to research the human heart, you will follow the above steps:

1. Write a sentence stating what you want to research.

   I want to find out the parts of the human heart and the function of the parts.

2. Identify keywords. Underline those in your sentence.

I want to find out the <u>parts</u> of the <u>human heart</u> and the <u>function</u> of the <u>parts</u>.

3. List synonyms of keywords.

parts = components, members, subdivisions

human heart = cardiac organ, vascular organ

function = capacity

4. Combine synonyms and keywords with or/and/not/+/–. Put parentheses around "or" phrases.

(parts or components or members or subdivisions) and

(human heart or cardiac organ or vascular organ) and

(function or capacity)

5. Check spelling.

EVALUATING WEB MATERIAL

In the past, the experts in their fields evaluated most reference material. Written science material such as found in journal articles, reference materials, and books are scrutinized by an editorial process that includes a review by science experts. Even films, CD-ROMs, and audiovisual material are reviewed by a standardized process before being made available to the public. In addition, the librarians evaluated most materials found in academic libraries. However, there is no standard evaluation practice in place for review of Internet material. You, the Internet user, are the evaluator. Since there is no evaluative process and since the Internet is accessible to anyone for uploading information, evaluating each Web site for accuracy, content completeness, and up-to-dateness is a critical part of the researcher's job. The following guidelines/questions will help you with this task:

- Who is the author? Is the author/organization one you recognize or one who others recognize positively? Is the author's resume included? Can you contact the author?

- Who is the publisher? Has the document gone through peer review?

- What is the author's bias? Is the article based on logic or logically obtained data?

- Is the article referenced? Are the references/links current? Do the links work? Is the site updated? Does the site list the date of the last time updated?

- If educational, is the material at the readability level stated?

Table 1.3
Web Site Evaluation Form

Name of Web site   _____
URL   _____
Web site author   _____
Date of last update   _____

Web site contains: Tutorial   _____
                 Laboratory activities   _____
                 Resources   _____
                 Professional information   _____
                 Teacher information   _____
                 General science information   _____

	1 (poor)	2	3 (average)	4	5 (excellent)
1. Article is referenced					
2. References are current (within 5 years)					
3. Author's résumé or e-mail address included					
4. Article is based on research and/or data analysis					
5. Links are current and operational					
6. Article is peer reviewed					
7. Navigation is easy					
8. Printability (easy to print, clear design)					
9. Site is interactive					
10. Updates are listed					
11. Content information—site is easy to follow					
12. Appropriate readability level					

In addition, you should evaluate the site for ease of use, technical availability, and for timeliness. In order to make this process easier, I have developed the checklist in Table 1.3 for your use.

BEYOND THE WEB

Other sources of information not on the World Wide Web should be mentioned. These resources include WAIS (Wide Area Information System), Gopher (an older, menu-based information system), Telnet (online library services), FTP (a comprehensive listing of anonymous sites), Experimental Metamap, and Experimental Search Engine Meta-Index. An interesting site for science is the White House Papers (WAIS).

COPYRIGHT AND PLAGIARISM IN THE ELECTRONIC AGE

Information is now readily accessible and at your fingertips. After you have evaluated a Web site's credibility and usability, there are rules governing the use of this information. Information concerning Internet materials usage vary from source to source. According to one Internet source, "ideas, facts, titles, names, short phrases, and blank forms cannot be protected by copyright." From a safe user position, (use also http://www.cs.csubak.edu, the Cyberspace Law for Non-lawyers Web site) consider everything that is written to be eligible for copyright. Therefore, if it has a copyright (which conservatively we assume everything written has) and you have decided to make a copy and you did not get the permission of the author, you may have violated copyright law. As in life, there are exceptions to the rules, called the Fair Use Doctrine. If you copy a piece and can answer "yes" to most of the following, you are allowed to copy legally as long as you cite the author:

- Is the piece for noncommercial use (you are not going to sell it)?
- Is the piece for criticism, comment, parody, news reporting, teaching, scholarship, or research use?
- Is your use of the piece mostly factual?
- Is the work intended specifically to be a reference work?
- Are you using only a small part of work?
- Are you using only an insignificant part of work?
- Are you adding significantly to original work?
- Does your use affect any profits that the original owner makes?

Remember: It's plagiarism if you do *not* acknowledge the author and if you use writings *unchanged*.

Use the following Web sites to clarify style and additional guidance:

Online! Citation Styles
http://www.bedfordstmartins.com/online/citex.html
A guide for Modern Language Association (MLA), American Psychological Association (APA), and *Chicago Manual of Style*/Turabian styles.

Electronic Styles: A Handbook for Citing Electronic Information
http://www.uvm.edu/~ncrane/estyles/apa.html
http://www.uvm.edu/~ncrane/estyles/mla.html
Style guide for electronic material.

Library of Congress: Citing Electronic Sources
http://lcweb2.loc.gov/ammem/ndlpedu/cite.html
Provides examples of citations for MLA and *Chicago*/Turabian style.

The Columbia Guide to Online Style
http://www.columbia.edu/cu/cup/cgos/idx_basic.html
A guide for MLA, *Chicago*/Turabian, and APA style.

2
RESOURCES IN EARTH SCIENCE

Chapter 2 is devoted to the Internet and specific Web sites. Specifically it is divided into three sections: general science information, including reference, commercial, and government sites, academic and educational sites, existing programs and demonstration sites; Earth Science information, including glossaries and dictionaries, tables and charts and formulas; an alphabetical listing of Earth Science topics.

TYPES OF GENERAL SCIENCE RESOURCES

This section contains an alphabetical list of general science resources. Each resource provides the URL, grade level, best search engine, key search words used, and a brief review of the site.

Reference

Title: CNN Science and Technology Page
URL: http://www.cnn.com/tech
Grade Level: All grade levels
Best Search Engine: http://www.metacrawler.com
Key Search Words: biology resources
Review: This site contains current and archived science news. The site has links to videos and audio archives. Type the name of your topic or subject in the search window provided on the home page. You can further refine your search in another search window. The resulting article summaries can be sorted by date or by relevance. If you sort by relevance, the key search word will be highlighted in yellow in the

summary and the article summaries will be ranked according to relevance. Clicking on the article link leads you to the complete article as well as links to related stories and to related sites.

Title: ERIC Clearinghouse for Science, Mathematics, and Environmental Education
URL: http://www.ericse.org
Grade Level: All grade levels
Best Search Engine: http://www.metacrawler.com
Key Search Words: science resources
Review: A component of the Educational Resources Information Center sponsored by the U.S. Department of Education. This resource was designed to provide information about learning science and the environment. It has links to science and environmental science resources, resources for parents and children, a bookstore, online publications, a conference calendar, and other science links.

Title: Electronic Journal Index
URL: http://www.coalliance.org/ejournal
Grade Level: High school–college
Best Search Engine: www.metacrawler.com
Key Search Words: educators sites + science + journals
Review: This Web site contains journals, magazines, newsletters, and other publications available over the Internet. The site lists the "electronic journals" alphabetically, by subject, by publisher, or by publisher in Colorado. The site contains a search window. When you use the search window option, the site provides you with a list of matches with a star indicator that ranks the listed Web sites as to appropriateness. For instance, a four-star ranking indicates a better match. When you click on the alphabetical or subject links, you reach a screen where you can click on an alphabet index (a, b, c . . .) or type into a search window. Once into a specific electronic journal, a complete address, peer review status, and subject list index are available. This is an excellent resource for you to begin your search for information.

Title: Enews: The Ultimate Magazine Site
URL: http://www.enews.com
Grade Level: All grade levels
Best Search Engine: www.metacrawler.com
Key Search Words: science magazines

Review: This site provides information concerning more than 900 magazines. You can search by title, category, interest area, or alphabetical listing. A brief review of the magazine, including subject area and grade appropriateness, is included. Access to information about purchasing popular science magazines, including *Discover, Popular Science, National Geographic World, Natural History, Audubon, Science News, Mother Earth News, Animals,* and *American Heritage of Invention and Technology* is also provided.

Title: LA Science Page
URL: http://www.latimes.com/news/science/science
Grade Level: All grade levels
Best Search Engine: www.metacrawler.com
Key Search Words: science resources
Review: The home page of this site provides a link to the archive depository. Clicking on the archive link leads you to a search window and to other search tools. You can type a topic into the search window, and further narrow your search to a specific period or a specific section of the paper. Two additional search tools help you narrow your search by relevance, by publication date, and by amount of stories retrieved. This service gives you access to all the articles on a specific topic published in the *Los Angeles Times* since 1990. Archive stories older than 14 days cost $2/story or $6/month for up to 10 stories.

Title: Nature
URL: http://www.nature.com
Grade Level: All grade levels
Best Search Engine: www.metacrawler.com
Key Search Words: biology reference
Review: Access to *Nature* magazine. Includes a trial, online copy of the magazine. Has a search icon, but you must register and pay to access. Has a product link to new products advertised in *Nature*.

Title: Science
URL: http://www.sciencemag.org
Grade Level: Upper grade levels
Best Search Engine: www.metacrawler.com
Key Search Words: science reference
Review: This site has an alphabetical science term listing, which you access through the browse link. Clicking on the browse link leads you to a long listing of articles on specific subjects. Each article is briefly

described, and links to the complete article or abstract are provided for each article. Therefore, you can print directly off the Internet, or you can access the Order an article or issue link and order for $8 per article/issue.

Title: The Philadelphia Inquirer Health & Science Magazine
URL: http://sln.fi.edu/inquirer/inquirer.html
Grade Level: All grade levels
Best Search Engine: www.metacrawler.com
Key Search Words: science magazines
Review: This site contains articles with teacher/student resources on the most current science developments. The Franklin Institute Museum maintains the site and offers materials for students/teachers.

Title: The Why Files
URL: http://whyfiles.news.wisc.edu/oldstorylist.html
Grade Level: All grade levels
Review: The University of Wisconsin maintains this site. You can navigate through the Web site by using a topic index (biology, physical science, etc.) or by using the provided subject search window. Basically, the site tries to answer the question "why?"

Commercial and Government Sites

Title: National Science Foundation (NSF)
URL: http://www.nsf.gov
Grade Level: College and above
Review: Congress established the National Science Foundation on May 10, 1950, in order to "promote the progress of science, to advance the national health, prosperity and welfare, to secure the national defense, and for other purposes." At the home page, you can link into the science programs or science highlights, or search by word/phrase or specialization area. Clicking on Biology at the home page takes you to the Biology program area. You can find information about special initiatives, funding opportunities, and news. There are also links to more specific biology disciplines such as Biological Infrastructure and Environmental Biology.

Academic and Educational Sites

Title: American Association for the Advancement of Science (AAAS)
URL: http://www.aaas.org

Grade Level: All grade levels

Review: The American Association for the Advancement of Science was founded in 1848 in Philadelphia. It is the largest science organization and publishes the peer-reviewed journal *Science*. At the home page, you are invited to explore more about AAAS, browse the online products, access the latest data of science and society, study the science educational programs for the future, or explore careers in science. The "on-line products" is the link to exploring present and past research on various topics. There are three levels of data retrieval, two of which require you to pay a fee. However, after registering for free, you can retrieve for free: full-text of articles published in the past year, staff-written summaries of research, abstracts of current or new research, access to the table of contents in all back issues, science search by author and keyword, and other resources too numerous to mention.

Demonstration Site

Title: Twinkies Project

URL: http://www.twinkiesproject.com

Grade Level: All grade levels

Best Search Engine: www.metacrawler.com

Key Search Words: science projects

Review: This is the most humorous yet well-researched and well-written site on scientific method and research. Twinkies stands for Tests With Inorganic Noxious Kakes In Extreme Situations. All of these "tests" were done at Rice University during finals week, obviously after much time spent studying for the normal final exams. At the main menu, you are led to a number of different "tests" that you can do with a Twinkie as your "chemical."

GENERAL SOURCES OF EARTH SCIENCE INFORMATION

Safety

Title: Occupational Safety and Health Administration

URL: http://www.osha.gov

Grade Level: College and above

Review: OSHA's mission is "to save lives, prevent injuries, and protect the health of America's workers." The OSHA home page presents you with either an alphabetical search tool or a search window to type in your search words. Clicking on the "M" leads you to a list of services and materials, including the "Material Safety Data Sheet," which is a must for all biologists. These sheets provide valuable information about every chemical. Unfortunately, more information is gotten directly from the OSHA site than is generally needed. It will take you time to download the MSDS guidelines and information for even one chemical.

Glossaries and Dictionaries

Title: YourDictionary.Com
URL: http://www.yourdictionary.com
Grade Level: All grade levels
Review: This is an online dictionary Web site. At the home page you can type your word in either the dictionary or thesaurus search window. On the left of the home page, you will find a list of specialty dictionaries; click on the "80 more!" and have other specialty dictionaries appear on the Web page. Clicking on the "biology" link leads you to a number of specific biology dictionaries and glossaries, including links to an "on-line biology dictionary" and "on-line biology glossary."

Formulas

Title: Wilton High School Conversion Formulas
URL: http://www.chemistrycoach.com/conversi.htm#
     ConversionFormulas
Grade Level: High school and above
Review: This Web page provides direct access to conversion formulas for length, area, mass, time, force, energy, momentum, power, number, volume, density, velocity, pressure, acceleration, frequency, angles, action, charge, current, resistance, EMF, and mass to force. This would be a very usable conversion table.

Title: Fundamental Physical Constants
URL: http://www.chemie.fu-berlin.de/chemistry/general/constants_
     en.html
Grade Level: High school and above
Review: This page was developed by the biology, chemistry, and pharmacy departments of Freie Universität Berlin. It contains an alphabetical list of physical constants, their symbols, and their numeric values with appropriate units. Including Avogadro's number, Boltzmann constant, and Planck's constant to name a few.

EARTH SCIENCE TOPICS

This section is designed for you to easily use. It is developed alphabetically according to Earth Science topic. Additional cross-listed topics will be included in the appendix for ease of use. Each review will contain the name of the Web site, URL address, appropriate grade level, search engine used, key search words used, and a review of the site.

Air Movements

Title: Air Movements
URL: http://www.aircurrents.org/about.htm
Grade Level: Middle to high school
Best Search Engine: www.metacrawler.com
Key Search Words: wind and air currents
Review: This is a site specifically designed for middle and high school students to get involved with and learn about air and air quality. The site is cleanly laid out, with the introductory page giving a concise overview of the project, its goal, and the process. It identifies how to participate, the "what" of the project and its needs, and funding that supported the development of the curriculum. This page also links to other sites either for information in general about pollution (epa.gov) or for more specific information about supporters of the project. The top of the page has buttons that send the viewer to other parts of the site with more specific information about participating in the Air Currents Project. One page is very specific about how to gather information, record it, and report it. Another explains how to post information and retrieve data from other participants in the program. The last of the buttons is the resource page with a list of Internet sites that provide additional or related information. This is a rather dry site intended to give specific and detailed information to those who are involved in the Air Currents Project. Because it is geared for students, it is written in plain English and gives comprehensive information. Good site and interesting project to manage over time.

Title: Movement of Air
URL: http://www.doc.mmu.ac.uk/aric/eae/index.html
Grade Level: All ages
Best Search Engine: www.google.com
Key Search Words: what causes air movement?
Review: Simpsons fans will get a kick out of this page. The *Encyclopedia of the Atmospheric Environment* created this easy to navigate site with

both young and older readers in mind. This site is available in English and French. By clicking on the English link, readers are taken to an introduction page. This page has nine main links: acid rain, air quality, atmosphere, climate, climate changes, global warming, ozone depletion, sustainability, and weather. Each link contains a host of information concerning a wide variety of environmental topics. For example, within in the atmosphere link, 30+ links are gathered on various topics associated with the atmosphere. These links are listed down the right side of the page. Each link is labeled with a Bart Simpson icon for younger readers or a Mr. Burns icon for a more detailed reading. Readers can choose if they want to click on Bart or on Mr. Burns depending upon what reading level they are interested in viewing. By clicking on Mr. Burns in the weather link, air movement could be located from the introduction to weather icon. This link provides information on air movement with links for various vocabulary words within the text. When these links were clicked, information about the specific words was presented. The reader could easily go back to the air movement page by clicking the Back button found on top of the computer screen. The only improvement that this site could benefit from is a search function because it offers a vast amount of information on various topics. Using www.google.com made it easy to find information on air movement. It is not clear how easily located the information would have been if the search engine was not used.

Title: Movement of Air
URL: http://www.doc.mmu.ac.uk/aric/eae/Weather/Older/Movement_ of_Air.html
Grade Level: Fourth grade and up
Best Search Engine: www.excite.com
Key Search Words: air movement wind
Review: Rather surprising to me was the fact that the Internet lacked any really useful information on air movement. In this technological age, I'd assumed that every topic has been explored from top to bottom and given a Web site devoted to it. Most Web sites that resulted from my searches were trying to sell me some kind of wind sock or other meteorological doodads. However, I finally stumbled on this small but informative page about the Movement of Air. This site thoroughly explains air movement and makes it quite clear and easy to follow. It introduces terms related to air movement such as temperature, pressure, wind, convection, uplift of air, fronts, and so forth.

Navigating through the rest of the site leads us on an adventurous tour through the world of weather. Many subtopics of weather are explored on this site, and there are myriad links for the always eager weather watcher. Again, though air movement was not explored in great depth, this site does serve as a great introduction and definition to air movement.

Title: Wind Direction and Speed systems by Wind Dancer
URL: http://www.winddirection.com
Grade Level: Fifth grade to adult
Best Search Engine: www.metacrawler.com
Key Search Words: air movement wind
Review: Finding a Web site for the topic air movements was very difficult. Many of the sites that I came across were advertisements and had nothing to do with air movement. Finally, I decided to give one of the advertisements a try and I clicked on the www.winddirection.com's home page. This site advertises the Wind Dancer, which is a new patented product that claims to measure wind direction and wind speed. It is an aerodynamically designed pennant with a pivoted harness system to animate the wind. It claims to be able to determine wind speeds as low as .05 mile per hour and up to one-quarter mile away. Also, it claims that this is a better device than the wind sock most people currently use. Personally, I am not convinced that this device can give someone a very accurate reading of wind speed; however, I'm sure it will provide a vague idea of wind direction. The Web site's home page provides bolded sections that you can click on to learn more about this product. For example, a section called "How does it work?" explains the product in detail with pictures of the design and wind charts. In my science class, I would have an activity where my students would try to construct one of these Wind Dancers and test it to see if it really works. This Web site would be very helpful for such an activity.

Title: Blizzard and Snow Theme
URL: http://www.cln.org/themes/blizzards.html
Grade Level: All ages
Best Search Engine: www.metacrawler.com
Key Search Word: blizzard
Review: This site provides the visitor with collections of Internet educational resources for blizzards and snow and other natural disasters. The site contains information for teachers as well as students. Within the

pages the visitor will find curricular resources as well as links to in-structional materials, such as lesson plans. The home page allows the visitor to access the Antarctic Theme Page, Arctic Theme Page, or Glacier Theme Page right off the bat. A visitor accessing the "Arctic Page" will be taken to a list of links to access all types of information about the arctic such as dozens of links that give insight about the icy world and its inhabitants. The Arctic Wildlife Portfolio is divided into an Arctic wildlife glossary, native birds, and mammals as well as facts about sea mammals. Another link I found particularly interesting was the "Biome Investigation" link. Here students use science process and research skills to study the six biomes of the world. The activities are designed with visual resources in mind. The link to the *Nunatsiag News* is a fantastic way for students to access and read about people a world away. This site also allows access to information on natural di-sasters such as avalanches, earthquakes, floods, hurricanes, tornadoes, tsunamis, and volcanoes. I chose to examine the link on "volcanoes. In using this link I was able to view locations, updates and images of active volcanoes by geographic region as well as five hands-on activi-ties that provide ways for students and teachers to explore volcanoes. The link to the Global Volcanism Program allows the visitor to check out the Smithsonian Institution's monthly *Bulletin of the Global Vol-canism Network and Archives*. The "Blizzard Attack" gave some really fantastic interactive lesson plans for grades 7–12 in which students journey between two cities during adverse weather conditions to ac-quire some basic skills necessary to survive winter storms. "Building an Igloo" can satisfy the interest of those who have always wondered how an igloo could possibly keep a person warm. This site also explains how to construct your own igloo (just in case you're stranded). The link "In Praise of Snow" is specifically designed for high school stu-dents. The link allows access to *The Atlantic Monthly*, a news bulletin that helps individuals watch weather conditions, understand them and even learn to predict weather through thorough articles. The "All About Snow" link has a beautiful gallery containing photos of historic blizzards. This link also provides simple explanations to questions such as Why is snow white? The link to "Mad Scientist Network" is a search engine in which the visitor can enter keyword terms to perform a search on not only blizzards but on any scientific topic of interest. This Web site was chock full of activities. They included such things as simple ways to construct snowflakes to teaching physical education in adverse weather conditions, such as snow. I highly recommend this site to teachers and students alike.

Air Pressure

Title: It's a Breeze—How Air Pressure Affects You
URL: http://www.kids.earth.nasa.gov/archive/air_pressure/index.html
Grade Level: All grade levels
Best Search Engine: www.metacrawler.com
Key Search Words: air pressure
Review: This page, part of a site created by NASA with a primary focus
   on information and education of children, is well organized, unclut-
   tered, and very user friendly. Throughout the page are text links that
   take the user to a glossary that defines words that may be unfamiliar
   to some children. The page begins with a clear, concise explanation
   of air pressure and the different ways one might experience it. There
   are then additional links for further explanation of air pressure and
   how it can be used in assisting weather forecasters and in making
   storm predictions, as well as an additional page detailing how one can
   make a barometer using simple materials. There are also brief descriptions of a variety of fun, interesting experiments to test and/or
   demonstrate air pressure; a link takes the user to another page that
   explains the experiment and its results. There is also a link to another
   page, which has a word search for the user to test their skills and vocabulary. There is another experiment in which the user is asked to
   take a guess at the results. The interactive experiment uses a hot air
   balloon and provides a demonstration and explanation of what is
   happening. At the bottom of the page, there are a number of discussion questions that students can take a crack at as well as one final
   question and explanation of air pressure and its effects in another sit-
   uation (this too has an interactive activity, but this time it is right on
   the home page). The site is full of interesting experiments and results
   to not only explain what air pressure is, but to also demonstrate it in
   a variety of ways. The pages are simple, clean, and very well laid out
   for easy navigation. All of the material is written in a concise manner,
   but without compromising the quality of the information. I think
   children would really enjoy this site and could learn a great deal
   about the main influences of air pressure in a very short time.

Title: What Happens When a Storm Comes?
URL: http://www.miamisci.org/hurricane/weatherstation.html
Grade Level: Grades 3–5
Best Search Engine: www.yahoo.com
Key Search Words: air pressure, elementary education

Review: Out of all of the sites reviewed, this site was the one that was geared to younger children rather than high school or college students. This Web site, created by The Miami Museum of Science, provides students with a definition of air pressure and many experiments to let the students obtain a hands-on feel of how air pressure is measured. There are eight topics that the students can choose from: Air Pressure, Conditions, Moisture, Project Materials, Temperature, Tools, Umbrella, and Wind. If the student clicks on any of the links, he or she will get easily understood definitions of certain weather terms. Each subheading contain directions to experiments, such as making a barometer, measuring wind speed, checking the power of air pressure, and reading thermometers. The directions are very easy to follow, and the tools needed to complete the experiments are simple to obtain. This site also links to Hurricane: Storm Science. It explains how measuring air pressure helps meteorologists predict storms. It also goes into a lot of information about hurricanes in general. There is also a link to the Science Learning Network, where there are 33 links to different science pages. Not all of the links pertain to air pressure or weather, but if the student had to do a science report in the future then this page may be a good place to start the research. This is a good site for children between the ages of 8 and 10 because it teaches them about air pressure in a fun way. There are many cartoons and illustrations of the experiments. The site is also easily navigated. It is a great site for teachers seeking ideas for class projects also.

Title: Understanding Air Pressure
URL: http://www.usatoday.com/weather/wbarocx.htm
Grade Level: All grade levels
Best Search Engine: www.metacrawler.com
Key Search Words: air pressure
Review: Understanding Air Pressure is from the *USA Today* newspaper. On the left side of the page is a menu that you can click on to go to other pages of the site. On the main area of the Web site is the introduction to air pressure. Following the introduction are the units of pressure, pressure, connects, and how pressure connects with other things. The last section is how air pressure decreases with altitude. At the bottom of the page are links to other sections of the site. The sections include the standard atmospheric table, density altitude explained, weather explained, weather calculations, weather works, and weather fronts. I liked this site because it came from a newspaper and it was well written.

Title: Observing Pressure
URL: http://www.miamisci.org/hurricane/airpressure.html
Grade Level: Grades 4–8
Best Search Engine: www.excite.com
Key Search Words: air pressure
Review: This site, developed by the Miami Museum of Science, provides
   basic information for its reader concerning air pressure. The page
   opens with a basic definition of air pressure. Questions are raised for
   readers as they scroll down concerning the measurement of air pressure, specifically as it relates to weather. There are two boxes that link
   to experiments that may be performed in the home or classroom. The
   first of these experiments uses a barometer to measure the air pressure. Students are instructed on how to build this barometer using
   common household items. The second box links to an experiment
   that examines the strength of air pressure. Again, using household
   items for materials, a step-by-step plan covers construction. This experiment involves a hot plate, which requires adult supervision.
   Below these two experiment links, the reader is presented with three
   more boxes that link to related topic areas. The first is labeled
   Weather Station and links the reader to the supplies and procedures
   necessary in creating a home or classroom weather center. Tools such
   as an anemometer, wind scale, wind chimes, wind streamer, and a
   rain gauge are introduced in this section. Procedures on both the construction and use of these tools are provided. The second box is labeled Hurricanes and links to a colorful map with several options for
   the reader to click on. Subtopics include Inside a Hurricane, Sur-
   vivors, Killer Storms, and Weather Instruments. In this section, the
   reader can look inside a real hurricane and read accounts from a family who witnessed a hurricane in their own home. The third box links
   to the Miami Museum of Science's home page. Here, the reader can
   view information on exhibits and programs operated by the museum.
   A link is also provided to the museum's online store, where it is possible to purchase CD-ROM versions of the experiments/lessons
   presented on this Web site. The CD-ROM versions make this information accessible to those who are unable to access an Internet connection. Overall, this is a very good introductory site for students
   concerning air pressure. The colorful and child-oriented graphics
   keep the attention of its readers. The vocabulary is age appropriate
   for the upper elementary/middle grades. Adult/teacher supervision
   may be necessary in explaining topic-related vocabulary and in performing the described experiments. The experiments introduced

would be valuable for a teacher to perform with the class as a whole or in small groups. This site provides both information and resources that can be helpful in teaching this unit to students using a hands-on approach.

Title: Science Fairs
URL: http://www.stemnet.nf.ca/sciencefairs/elem.html
Grade Level: All grade levels
Best Search Engine: www.metacrawler.com
Key Search Words: air pressure elementary
Review: This site is wonderful! It gives pages full of possible science projects for elementary and high school students to do, not just with air pressure but with many interesting topics. On the left of the home page are various buttons, all clearly labeled, that one can click on for more information. The Primary projects box gives ideas for first through fourth grades, to make display-type projects, and the tip at the top tells the teacher it is best for students to really try to make their own observations, rather than simply copy something out of a book and not learn anything! The Elementary projects button lists both display and experimental type project ideas for grades four to six, where they can build various models and conduct experiments. The Intermediate projects box gives seventh- through ninth-grade project ideas, built around topics like Physics, Engineering, Meteorology, Chemistry, and Botany. The Senior projects page is for high schoolers and is centered on Biology, Engineering, Physical Science, Environmental Science, Meteorology, and Computer Science. There's also a box that links you to other science Web sites if you need additional help, and there's even a box for questions and comments, so you can get clarification on something or give the site's developers a project idea yourself. The site was easy to navigate and easy to understand, and it gave me countless ideas for future projects; you could use only this site and have enough material for a whole school year! It is more of a site for teachers, though, than students, at least for the younger grades.

Title: Reeko's Mad Scientist Lab
URL: http://www.spartechsoftware.com/reeko/Experiments/
      magiccan.htm
Grade Level: Middle and high school
Best Search Engine: www.dogpile.com
Key Search Words: air pressure

Review: This provides students with an experiment demonstrating and explaining the concept of air pressure using a coffee can. The materials needed for this experiment are provided at the top of the page. Before the procedures of the experiment are given to the students, an explanation of air pressure is provided. After performing this experiment, one will fully understand what air pressure is and the concept behind it. The site provides a glossary defining numerous amounts of words pertaining to air pressure and science. Additional experiments relating to sound, light, and motion are also available. This site provides laboratory safety rules for the students to use when performing any type of scientific experiment. All in all, this site is an excellent resource.

Title: Air Pressure
URL: http://www.antarctica.ac.uk/Operations
Grade Level: 9th grade and higher
Best Search Engine: www.google.com
Key Search Words: air pressure
Review: The site contains information for people traveling to and from the Antarctic with the British Antarctic Survey (BAS). The eight buttons on top of the site are About BAS, Key Topics, BAS Science, Living and Working, About Antarctica, News and Info, Resources, and Search. These buttons take you to informative information about the site and Antarctica. For example, About Antarctica gives you history. Other links go to Weather, Wildlife, Ice, Rock, Antarctic Islands, Tourism, Music, Environment Management and Conservation, British Antarctic Territory, and the Antarctic Treaty. The Antarctic Treaty button takes you into the countries involved in the treaty and the five international agreements. Back to the home page: The center of the page provides a brief scientific background of BAS. In the middle of the page there is a link: Science Programme. When you click on this link it gives you the programs from 2000 to 2005. In addition, there are signals of Antarctic past and global changes. There were great pictures of the Earth. Other links on this page include Global Interaction of the Antarctic Ice Sheet and Antarctic Climate Process. Located on the left side of the site were links to additional information. All of links take users to Projects, Challenge, Objectives and Delivering the Science. Located on the left side of the screen: Magnetic Reconnection Substorms and Their Consequences, Geospace Atmosphere Transform Functions, Antarctica in Dynamic Global Plate System, Antarctica Biodiversity Past, Present and Fu-

ture, Life at the Edge-Stresses and Thresholds, Dynamics and Management of Ocean Ecosystems, Independent Projects, and Environmental Research, and Medical Research. This site was very scientific. It had beautiful photos and projects. It is designed for mature students or adults. However, younger students would be interested in the wildlife and rocks in Antarctica.

Title: Air Pressure and Humidity
URL: http://www.factmonster.com/ipka/A0769510.html
Grade Level: Grades 3–5
Best Search Engine: www.google.com
Key Search Words: air pressure
Review: This site, sponsored by the Learning Network, is fun and educational for children. The Fact Monster home page is colorful and contains interesting graphics that serve as the links to the following: World and News, U.S., People, Word Wise, Math, Science, Sports, Cool Stuff, Games & Quizzes, and Homework Center. The home page also displays a column of daily features. This column offers a word quiz and a geography guide, as well as other poignant issues for the day. By clicking on the Science icon you are taken to the science home page, which contains three sections, Almanac, Special Features, and Games & Quizzes. In the almanac section you will find links to the following: Environment, Body, Animals, Stars & Planets, Weather, Food, General Science, and Computers & Internet. By clicking on any of the links above, you are taken to another page with at least 20 topics about the link you have chosen. For example, when clicking on the weather icon you are directed to a page with many weather-related topics. Here, you can find a link to information on air pressure and humidity. The material is laid out in an organized manner. The points are bulleted and the wording is concise and easy to understand. The second section on the page, called special features, serves as a daily topics board. Here, you can find many interesting facts about the subject of the day; on this day it happened to be sharks. Finally, the third section on the page consists of fun games and quizzes on the information offered within the site. I think that this site is fantastic for children in grades 3–5. The information is factual and relevant, while the site's design is inviting for kids. Most importantly, the site is organized and easily navigated. I think www.factmonster.com is a great resource for kids and adults.

Air Temperature

Title: Chapter 3: Air Temperature
URL: http://www2.una.edu/geography/classes/ge101/101ch3.html
Grade Level: High school
Best Search Engine: www.metacrawler.com
Key Search Words: air temperature
Review: This is a very informative Web site on air temperature. The information comes from a science textbook. Air temperature is reviewed and described in great detail to the reader. Words pertaining to air temperature such as conduction, evaporation, and thermometer are defined in detail. Topics pertaining to air temperature such as the measurement of air temperature, the daily cycle of air temperature, the temperature structure of the atmosphere, the annual cycle of air temperature, global warming, and the greenhouse effect are discussed in the chapter on air temperature. Each topic is covered in great detail, containing subtopics to the main topics. Because the information is from a textbook, it is written on a level that a student will understand. All in all this site is an excellent resource for detailed information about air temperature.

Title: Weather Basics
URL: www.usatoday.com/weather/basics/wworks0.htm
Grade Level: High school
Best Search Engine: www.yahoo.com
Key Search Words: air temperature and education
Review: There were not many sites dedicated completely to air temperature, but this site had a couple of articles on temperature, seasons, and heat. USA Today created the page, and it contains information on all aspects of weather, but you can find some stuff on air temperature. The main heading under which most of the info on air temperature is found is called "Learn How the Sun Drives Our Weather." Under the heading are eight articles to click on. For example if you click on "Air Masses and their Sources" you will find a brief description on air masses, plus an interactive graphic of the world that lets you see what type of air mass covers different regions. From that section you can click on a link to more details on different kinds of air masses. This site would be good to use if you had to research weather in general because there are about 150 articles written on everything to do with weather. I would not say this is a good site for elementary or middle

school students because they may get confused while navigating. But if an adult or high school student wants to get some quick, technological answers on air temperature, then this would be a good page to come to.

Title: About Temperature
URL: www.unidata.ucar.edu/staff/blynds/tmp.html
Grade Level: Middle and high school
Best Search Engine: www.metacrawler.com
Key Search Words: air temperature
Review: This site was developed by teachers involved in Project Sky-math. This project integrated science and math concepts very well. The home page has gold stars on the left side to designate each category. When you click on the star, there is detailed information on each topic. The first topic is "What is Temperature," which describes thermal equilibrium, which is when two objects of the same material are the same temperature. The next section is "The Development of Thermometers and Temperature Scales." This section includes formulas and graphs to illustrate how a thermometer works. There is a detailed explanation of how the Fahrenheit and Centigrade scales were developed, and there is a conversion formula. Throughout the text highlighted vocabulary words link to additional information such as primary reference points, liquid helium, and low temperature laboratories. The next section, "Heat and Thermodynamics," is more technical and explains the laws of thermodynamics. The next topic is "Kinetic Energy," which contains a lot of history on physicists and their discoveries in the field. There are formulas to explain each theory. The section on "Thermal Radiation" includes a diagram and an electromagnetic spectrum. The visuals help to understand the concept. The last section, "3K The Temperature of the Universe," discusses how the sun and stars emit thermal radiation covering many wavelengths. This site is informative, but a bit too technical. Although it was designed for middle school, the content would be suitable for high school students as well. The site is well organized and includes interesting facts about how theories were developed.

Title: Weather Basics
URL: http://www.usatoday.com/weather/basics/wworks0.htm
Grade Level: Grade 5 and up
Best Search Engine: www.yahoo.com
Key Search Words: air temperature and weather

Review: Trying to locate a Web site just on Air Temperature was very difficult. *USA Today* created this page to contain information about the weather, but you can find some information on Air Temperature. This site was very helpful when it came to Sun, Wind, Storms, Temperature, Seasons, and Heat. There are many headings such as Learn How the Sun Drives Weather. This heading has the most information on air temperature. This heading has 8 subheadings underneath it. The next heading is The What and Why of Wind with 16 subheadings. Storms and Fronts is next with 19 subheadings. Following is When Water Changes Forms with 15 subheadings. Floods and Droughts has 9. Snow, Cold, and Ice has 16. Lightning has 6. Thunderstorms has 20. Tornadoes has 9. Hurricanes has 18. The Sky has 8. Predicting the Weather has 5. Weather in Our Future has 12. Extreme Weather has 6. Finally, Books related to Weather has 4. Under the first heading if you click on "Air Masses and their sources," you will find a brief description on air masses. You can also see what type of air mass covers different parts of the world. From this section you can click on a link to receive more details on different kinds of air masses. This site would be very helpful if you wanted to research anything about weather. This site was easy to use and very helpful. I loved the way it breaks down each of the headings because it was very easy to get to where you want to go.

Title: Air Temperature Patterns
URL: http://www.uwsp.edu/geo/faculty/ritter/geog101/lectures/
    lecture_atmospheric_temperature.html
Grade Level: Science-literate, 11th grade to adult
Best Search Engine: www.google.com
Key Search Words: air temperature
Review: Navigating the Site: The site is an outline of a lecture titled Air Temperature Patterns for a college course, Geography 101. It contains three main sections: Introduction, Controls over air temperature at a place, and Global patterns of Air Temperature. The section on Controls over air temperature covers four topics: Radiation transfer and temperature, Air Temperature and sensible heat transfer, Air Temperature and water bodies, and Air Temperature and Air Mass Movement. Within the Radiation transfers and temperatures topic the bulleted item "Radiation, cloud cover and air temperature" links to an article in the Weather section of the *USA Today* site (www.usatoday.com/weather), "Why cloudy nights tend to be warmer." (See the reviews of *USA Today*'s site.) A final section, Can You . . . , poses

three review questions based on the information presented. The combination of an outline format and technical terminology would make this site difficult for students younger than eleventh grade and for adults with a limited familiarity with scientific terms related to this area. For example, I was surprised that this topic fell within a geography course, putting me firmly within the latter group. The graphics and links add interest to the site and help the reader understand the abbreviated outline-style text employed. This was the best site I could find on air temperature; in fact, MetaCrawler did not yield anything useful. However, because some of the terminology is not fully explained, I don't find this site ideal for someone who wants to acquire a good basic understanding of the topic.

Title: What's up with Air? Climate Change Causes and Effects
URL: http://thechalkboard.com/Corporations/Dow/Programs/
    EducatorGuide/Air/AirTOC.html
Grade Level: Middle school
Best Search Engine: www.google.com
Key Search Words: air temperature in the classroom
Review: This site, written from an educator's perspective, serves as a
    guide for conducting a science project about our climate and its air
    temperature. The goals of the project are outlined on the home page
    as follows: students will learn the components of "air," how its quality
    is affected by natural and human activity, and the effects of climate
    change on our lives. Students will learn air is an essential resource to
    be protected from pollution and we should reduce or restrict human
    activities that damage it. As you continue down the home page you
    will see links directing you to various aspects of the project. The first
    is a link for classroom specifications for the project. It includes information regarding the student skills that will be applied, student learning outcomes, parent and community involvement, time range, as
    well as the cross curricular learning outcomes for the project. I think
    the next link, teacher preparation, is most effective. Once clicked
    you are taken to an organizational tool and information page. The
    links include the following: vocabulary, materials and resources, and
    scheduling and planning. The information provided was very detailed and the vocabulary established a good foundation. As you
    scroll down this home page, tips for project management, lab work,
    and pre- and post-assessments, are available by choosing the various
    links. Furthermore, in the Extra! Extra! section, the site offers creative extension activities for you and your students to try. In addition,

this site provides teachers with a career awareness department. It offers many career choices for someone interested in the various weather occupations. I thought this site would be very useful for teachers in teaching about air temperature and the environment to middle school and junior high students. However, I found the information to be somewhat dry and the site's design certainly lacking.

Title: Ask a Scientist
URL: http://newton.dep.anl.gov/askasci/wea00/wea00061.htm
Grade Level: College
Best Search Engine: www.yahoo.com
Key Search Words: air temperature
Review: The air temperature Web site that I reviewed deals with dew points of the Earth. The site is from a weather archive. At the top of the site are two pictures. One picture is the logo for the Department of Energy, and the other picture informs the reader that this Web site comes from the University of Chicago. Underneath the pictures are questions and answers. Each question is answered by various people, and they all have a different answer. I did not like this site because there were no links. I also felt that it wasn't giving any sufficient information.

Title: About temperature
URL: http://www.unidata.ucar.edu/staff/blynds/tmp.html
Grade Level: Middle school
Best Search Engine: www.metacrawler.com
Key Search Words: elementary, air temperature
Review: The site was developed for middle school math teachers and is available in both English and Spanish! Different subtopics under the temperature home page can be accessed by clicking on the star icon next to them; there are topics such as the basic "What is temperature?" and topics like the more in-depth "Development of thermometers." The site gives the teacher a crash course, or perhaps a refresher, on temperature, thermometers, thermodynamics, the kinetic theory, thermal radiation, and much more. Several vocabulary words are linked to a detailed explanation, sometimes with pictures. It was easy to navigate around this Web site but difficult to understand the actual content of it.

Title: Air Temperature—Weather Maps
URL: http://cirrus.sprl.umich.edu/wxnet/maps.html

Grade Level: Grade 6 and up
Best Search Engine: www.lycos.com
Key Search Word: temperature
Review: I found it difficult to locate a Web site which was dedicated to air temperature. A few sites focused on sections of studies and/or lectures on topics related to air temperature. These sites appeared too advanced and intricate to be considered for student use. Many of the sites found were also professional sales sites, relaying information on equipment used in air temperature regulation. However, I did come across a site related to air temperature and the weather. This site, titled UM Weather, is a nonprofit service sponsored by The Weather Underground program within the University of Michigan at Ann Arbor. The main UM Weather site contains links to resources concerning weather phenomena and reports across the United States. Through my search on www.lycos.com, the Weather Maps link within this site was returned. The Weather Maps page displays three main links. The first link directs readers to information on National Surface Maps, the second Surface Temperatures, and the third Upper-air Maps. Maps of the United States are displayed to illustrate the current weather trends related to the chosen subtopic. The maps that the page links to successfully are of fair quality. The graphics/labels on these maps were not as clear as they could have been. Some of the links also did not function correctly, returning a "Page not Found" error message. This site could serve as a valuable resource for both teachers and students. However, it is not completely functional at this time and appears to require some further editing by its developers.

Title: Common Meteorological Variable
URL: http://k12.osc.edu/teachers
Grade Level: Grade 3 through adult
Best Search Engine: www.google.com
Key Search Words: science, air temperature
Review: I had a difficult time finding a site that interested me on air temperature, but I think this one does the trick. On this site, this page is listed under unit 2 under the references column. It is well organized and easy to read. The first section gives an explanation of air and its components. If you click on any of the "blue" words, you will be taken to a page of definitions. I thought this was great because you will not have to stop and look elsewhere for clarification of words that are unfamiliar. The next section on the page is an air temperature scale, which shows Fahrenheit, Celsius, and Kelvin. It also explains about

temperature and these measurements. Next is a section about dew-point temperature and relative humidity. The next illustration explains the relationship between air pressure and air temperature. The last section of this page explains the relationships between wind, air density, and water vapor mixing ratios. This site is easy to read and look at. I feel this site is very informative.

Title: Measuring Air Temperature
URL: http://www.nssl.noaa.gov/~cortinas/1014/l10_2.html
Grade Level: Grade 9 and up
Best Search Engine: www.google.com
Key Search Words: air temperature
Review: This page is linked to a university meteorology and weather course that investigates a number of weather topics. The page found at the address listed above deals only with measuring air temperature. It provides a clear, concise description of the types of instruments used to measure air temperature and what they are made up of as well as the proper ways in which to measure air temperature using that particular instrument. There is an illustrated example of a common thermometer and its different parts. There is a text link at the bottom of the page that leads to an introductory page, Human Comfort and Measuring Temperature. There are two main text links found here. One is the initial page that the search leads us to, Measuring Air Temperature, and a second page, Air Temperature and Human Comfort. This second link leads to a page, similar to the initial page, that provides a page of text that details the relationship between the temperature of the air and the comfort level of a human body. Other links found on the title page for the two detailed pages include Home, Syllabus, Lectures, Grades, and Links, making it obvious that this page is part of a university course. The pages are plain, yet they provide valuable and interesting information to the person seeking information on air temperature. In searching this topic, it was difficult to find sites and/or pages dedicated to air temperature alone. Most of the search results were very specific to things such as air temperature records or to the words *air* or *temperature*. This site at least provided practical, useful information on how to measure air temperature as well as an example of the relationship air temperature has to a human's comfort level. This page provides good, specific, and appropriate information related to air temperature measurement, but it is not an inviting or interesting site for the general user. For someone looking for specific information on measuring air temperature it would be a valuable resource.

Ancient Environments

Title: Ancient Environments
URL: http://www.fp.ucalgary.ca/pubarky/kid's_page.htm
Grade Level: All levels and teachers
Best Search Engine: www.yahoo.com
Key Search Words: ancient environments
Review: This was a very cool archaeology Web site set up by the Pro-
    gramme for Public Archaeology. Seven buttons are located on the top
    of the main page: Home, Public Excavations, Year Round Lab, Kids
    Page, Teachers, Eyes on Archaeology, and Ceremonies. These links are
    great for teachers. The Kids Page links to a page with lots of informa-
    tive links for kids to explore. One of them, Rock Art Images, lets kids
    examine rock images and submit their responses. This is a fun Web site
    for kids and teachers. Teachers are able to learn new teaching tech-
    niques. The site is easy to navigate, kid friendly and very colorful.

Title: Ancient Environments
URL: http://www.gns.cri.nz/earthhist/ancient_env/index.html
Grade Level: Grade 9 and up
Best Search Engine: www.google.com
Key Search Words: ancient environments
Review: This site is a subpage of a home page of the Institute of Geologi-
    cal and Nuclear Sciences Limited. This organization is sponsored by
    the government of New Zealand and calls itself an expert in geologi-
    cal and nuclear science. The subpage labeled Ancient Environments
    gives a brief description of how scientists gain information of past life
    on Earth, from fossil remains of plant and animals and sediment strat-
    ification. This page has four main links that cover the past 80 million
    years of geologic history in New Zealand. These links are typed in
    green and labeled Pleistocene glacial-interglacial cycles, Middle
    Miocene greenhouse-icehouse transition, The Late Paleocene Ther-
    mal Maximum, and The Cretaceous/Tertiary Mass Extinction. Each
    link provides concise information and citations of specific research
    on various areas of study, including the stratification of isotopes found
    in sediment and the chemistry and biology of fossil remains. Al-
    though this site is rather boring to look at, it provides sophisticated
    textual information on the study of ancient environments. One can
    easily toggle back and forth between pages because links to each of
    the four topics listed above can be found on any page. Three other
    icons on the Ancient Environments page allow you to visit the main

page, search for key words, and e-mail comments or suggestions for the site. The language that is used is pretty advanced, so this site is not recommended for young readers. Since this site contains links to citations of research projects, it would be a good research tool for high school or college students seeking background information on geology and as a springboard for finding primary sources to study.

Title: Inferring Ancient Environments from Fossil Foraminifera
URL: http://www.ucmp.berkely.edu/fosrec/olson3.html#FIG2
Grade Level: Grade 8 and up
Best Search Engine: www.metacrawler.com
Key Search Words: ancient environments
Review: In my search for Web sites that contained information about ancient environments, I found a site produced by Hilary Clement Olson. This site is most suitable for teachers of middle school. The site is straightforward and contains an activity called "Inferring Ancient Environments from Fossil Foraminifera." This site list prerequisites in order for the students to feel comfortable with the activity. Among the "prerequisites" list is the term "foraminifera." Here the visitor may click on "foraminifera," which links to a page that explains the term in depth and discusses how different benthic foraminifera prefer particular environments, which may be associated with water depth. While visiting the "Introduction to Foraminifera" site, the user may click on buttons to explore "Fossil Records," "History and Ecology," "Systematics," or "More on Morphology." Among the "prerequisites" list was the concepts of geologic time and petroleum reservoir rock and rock sources. In mentioning reservoir rock and source rock, the site offers a link "Miocene" which takes the visitor to a site titled "The Miocene Epoch." The site discusses that the Miocene was a time of warmer global climates than those in the earlier Oligocene period. The visitor may further explore "The Miocene Epoch" by clicking buttons listed "Stratigraphy," "Ancient Life," and "Tectonics." There is also a chart that shows the major subdivisions of the Neocene that include the Miocene. Any subdivision on the chart may be clicked to navigate the exhibit. Back to our original site, the activity lists the objective that explains that each student will use a reference diagram of fossil foraminifera with a paleo-water-depth assignment to interpret the water depth of a particular area of California during the geologic past. All step-by-step instructions are included along with nicely illustrated diagrams. Figure 1 illustrates species of foraminifera that lived at four different water depths. Figure

2 is a sample locality map, and Figure 3 is a sample analysis and interpretation chart of 10 types of foraminifera that were found in the four depth locations. Overall, I found the site easy to navigate and informative. If the teacher requires further information there is also a list of references.

Title: Ancient Environments
URL: http://www.gns.cri.nz/earthhist/ancient_env/index.html
Grade Level: Grade 6 and up
Best Search Engine: www.excite.com
Key Search Words: ancient environments
Review: After one glance, I was going to click the Back button on my browser and look for something else because it didn't seem like this site would have too much content. After doing a double take, I noticed that this site had much to offer. The first page is more or less an introduction to ancient environments and a statement on the aims of the researchers involved in the studies. This page also breaks down into the four episodes of environmental change that have occurred in the past 80 million years. These four periods are classified as: Pleistocene Glacial-Interglacial cycles, Middle Miocene Greenhouse-Icehouse Transition, The Late Paleocene Thermal Maximum, and The Cretaceous/Tertiary Mass Extinction. These periods are all links to more pages about each specific period. While I won't go far into depth about each period, the Web site certainly does. It gives a good overview on each period, citing how many million years ago it occurred, what conditions were like on the Earth, why these conditions were occurring, and so forth. Links are also given on each page to further your knowledge of each period. Overall, this is a very useful site for Earth Science and the exploration of ancient environments.

Title: Grand Canyon Ancient Environments
URL: http://www.edu-source.com/GCPages/CVOpage8.html
Grade Level: Grades 6–10
Best Search Engine: www.metacrawler.com
Key Search Words: ancient environments
Review: This is a simple and straightforward sight depicting the geological layers of the Grand Canyon as a picture of the development of the Earth. The main page gives a brief overview and history of the layers with good graphics and photos. This is part of a site on the natural history of northern Arizona. At the bottom of the page are links to pages that discuss the major layers of earth, the prehistoric environ-

ment that contributed to each layer, and the length of time the layer took to "build." The language is uncomplicated and written in an easy, informative style. While not an in-depth analysis of the environment, it does provide enough information for a report covering multiple topics or enough information to lead to further study. What is missing are the links to other sites; this one is self-contained within the master site. This is a good site for a quick overview about the history of the Grand Canyon and its geology. For more detailed information, the student would have to keep searching.

Title: Institute of Geological and Nuclear Science
URL: http://www.gns.cri.nz/earthhist/ancient_env
Grade Level: Middle school and up
Best Search Engine: www.msn.com
Key Search Words: ancient environments
Review: The Institute of Geological and Nuclear Science Web site provides a vast array of information on the Earth's environment, and particularly the ancient environments and how their changes affected the Earth. This institute is based in New Zealand, as the area offers the best Southern Hemisphere records for all four major episodes. Many countries, including the United States, benefit from the research that takes place here, and have research teams involved in this collaborative study. This site is user friendly with many sidebar items to get an in-depth discussion on broader topics of each page. At the top of the site there is another menu bar for more broad topic selections relating to the environment. You can select from Earth History, Earth Resources, and How Can We Help, among others. On the main page, it gives links to detailed information on the four major episodes of environmental change over the past 80 million years. It also describes briefly the indicators looked for when studying past environments. When you select the "What's New" icon in the top menu bar, there is a "kids" area where children can access an online seismograph and track recent earthquakes. The interactive portions help to get the children involved and gain a better understanding of what they are learning about.

Title: Royal Commission on the Ancient and Historical Monuments of Scotland
URL: http://www.rcahms.gov.uk
Grade Level: Grade 5 through adult
Best Search Engine: www.yahoo.com

Key Search Words: ancient + environments
Review: Clicking on "showcase" at the top of the page takes the user to
19 showcases of ancient and historical monuments of Scotland. Each
showcase offers brief descriptions and beautiful, enlargeable pictures.

Title: Ancient Environments
URL: http://www.wightonline.co.uk/wight/wight_pages/dinosaurs.html
Grade Level: Grades 2–6
Best Search Engine: www.google.com
Key Search Words: ancient environments
Review: At the top of the page it states: "Dinosaur Island—One of Eu-
rope's Finest Sites for Dinosaur Remains." This opening page is very
short and has one picture of the island. On the bottom of the page is
a link, Naturally Wight Contents. When you double click on that
link, you get several more: Island for all Seasons, Coloring the Island,
Exploring the Island, Coastal Wight, Rural Wight, Download Discovery, River Valleys, Forests and Woodlands, Secrets of the Under-
cliff, An Historic Landscape, Discovering Archaeology, Dinosaur
Island, The Garden Isle, and Areas of Outstanding Beauty. Each link
has a very short description and one unclear picture. The names of
the links are appealing, but they were disappointing. The site lacks
depth and facts. This site is geared toward a younger age level.

Atmosphere

Title: Learn: Atmospheric Science Explorers
URL: http://www.ucar.edu/Learn
Grade Level: Grades 5–8
Best Search Engine: www.metacrawler.com
Key Search Words: atmospheres + science + education
Review: Although I was sorely tempted to search for Web sites related to
extraterrestrial atmospheres (e.g., Jupiter, Saturn, and Mars), my attention was piqued by a site for teachers called "Learn: Atmospheric
Science Explorers," which covers the cycles of the Earth and atmosphere. The module overview (http://www.ucar.edu/Learn/1.htm) describes the site as an "on-line teaching module for middle school
science teachers . . . the primary audience is classroom teachers and it
has been developed with that target audience in mind. The site provides background information and supporting classroom-teaching
materials. The content focus is climate change and issues related to
both stratospheric and tropospheric ozone." There are seven sections

of the module that as a matter of course provide a list of activities and background information. The seven sections are Introduction to the Atmosphere, Introduction to Climate, The "Greenhouse Effect," Global Climate Change, Introduction to Ozone, Stratospheric Ozone, and Tropospheric Ozone. The module overview speaks on how "it would be among the most profound discoveries in the history of our species if we could detect life on other planets circling other stars in the galaxy." The authors of the site then pose a fascinating idea: "How planets block the light of their star as detected from Earth can reveal the chemical composition of the planet's atmosphere. . . . How is this possible? How can simple gases tell us about the presence of life as we know it?" The site goes on to explain about the biosphere and the concept of cycles; that certain materials and substances "move endlessly throughout the Earth's biosphere, atmosphere, and lithosphere, existing in different forms and being used by different organisms at different times, but always moving, always circulating." This section goes on about dynamic versus static equilibrium. "Introduction to the Atmosphere" provides a brief overview of the properties associated with the Earth's atmosphere including its composition, layering, how energy is transferred between the Earth's surface and atmosphere, and how ocean currents play a significant role in transferring this heat poleward. Activities follow each of the seven modules. For example, in the first module there are seven activities, including "The Goldilocks Principle—A model of Atmospheric Gases." The site author(s) note that "this activity introduces students to the atmospheric differences between the three "sister" planets in a graphic and hands-on way." As an added bonus, National Science Education Standards, Benchmarks for Science Literacy, Breakdown of Grade Level/Time, Materials and Procedures, Assessment Ideas, Modification for Alternative Learners, and additional resources are available for the teacher and the learner. Additionally, there are more than 25 lab activities spread over the seven major parts of the module, which means that teachers have the luxury of tailoring activities for different "sections of students." For example, "Introduction to Climate" discusses the general distinctions between weather and climate; speaks about long-term climate data, averages, and variability; discusses how the Earth's climate has changed over time; and finishes with a discussion of pollen and dendrochronology. In my favorite activity, named "Dinosaur Breath," students learn about the role of dinosaurs in the carbon cycle and the eventual storage of excess carbon in the form of chalk. For brevity's sake, I won't break down every section of the entire module, which would take another entire page and more, but leave it to say that every part of all seven modules are packed with artful drawings, illustrations, diagrams, and charts. The site provides detailed information and activities that seem ready to use, along with seemingly tried and tested material that appears useful with few if any changes needed, except for local availability of materials and equipment. While the site really "works," in terms of the background information, it is in the activities and all that goes with them that this site really shines. It provides fun, useful experiments and activities that seem as enjoyable to run for the students as they would be to perform as students.

Title: Exploration—Earth's Atmosphere
URL: http://liftoff.msfc.nasa.gov/academy/space/atmosphere.html
Grade Level: Grade 6 and up
Best Search Engine: www.google.com
Key Search Word: atmosphere
Review: This is a subpage of the Liftoff to Space Exploration site hosted by NASA. The first text link at the top of the page reads "Liftoff Home," and links to the site's home page, which has some additional links that are as terrific as the page being reviewed here. They include NASA Update, Spacecraft, Human Journey, The Universe, Fundamentals, and Tracking—all of the links are packed with great information and activities. The atmosphere page found at the above address is excellent. It provides detailed information about the Earth's atmosphere and its function, distance, and composition. It further details the four layers (troposphere, stratosphere, mesosphere, and thermosphere) of the atmosphere and provides information on each layer's distance and unique characteristics, including temperature and functions. There are a few text links located within the description paragraphs, but I think more would be useful because many technical terms are used on the page. Those text links that are provided go to pages that further explain the terms being described (consistent with the rest of the site, there is also excellent content found on these pages!). The bottom portion of the page has a section called Composition of the Atmosphere, which lists the components that make up the atmosphere, and a final section, Beyond the Atmosphere, which explains the exosphere. Overall, this is a clean, well organized, and informative site that would be interesting for students and teachers alike who want to learn about the atmosphere and its multiple layers

and functions. There is nothing on this page that is not interesting or useful to someone working to learn about the Earth's atmosphere.

Title: Atmosphere
URL: http://www.f.about.com/z/js/spr02.htm
Grade Level: Grade 6 and up
Best Search Engine: www.metacrawler.com
Key Search Word: atmosphere
Review: This Web site is very informational and has a lot of good links and helpful navigation hints. I would recommend this site to middle school and up. I think college students would find this site helpful for brushing up on atmosphere topics or definitions. If you click on the heading that states Project Atmosphere Australia Online, this will give you activities for teachers, photos, and great downloads. You can also click on Atmosphere and Climate Links, which is an intensive Web weather resource. This will give you a wide variety of atmosphere and climate topics. Under the heading The Earth's Atmosphere you can learn about the layers of the atmosphere and find out facts about the composition of the air we breathe. You can also click on Atmospheric-Ocean Science and find out how air and water work together to create and influence the climate. I enjoyed this Web site a lot and found it to be full of information.

Title: Why Is the Sky Blue?
URL: http://meto.umd.edu/~ezra/whyblue/whythmb.html
Grade Level: Middle and high school
Best Search Engine: www.dogpile.com
Key Search Word: atmosphere
Review: This Web site was created by Dr. Richard Berg from the University of Maryland. Dr. Berg discusses the atmosphere and tries to answer some of the most popular questions about the atmosphere, such as, Why is the sky blue?, Why can't you see the stars during the daytime?, and Why are sunrises and sunsets a reddish-orange color? Instead of using wordy explanations, Dr. Berg uses slide shows. There is a short but detailed explanation before each slide show, but the slide show does majority of the explaining to these questions. To see a slide show, one just has to click on the words "slide show" under each of the questions that they are interested in finding out the answer to. It is very interesting to see how someone fully answered these common questions using just pictures. Students who are visual learners will

find this a useful site. All in all, this site is an excellent resource for better understanding the atmosphere.

Title: ECOworld The Global Environment Community
URL: http://www.ecoworld.com
Grade Level: Grade 3 to adult
Best Search Engine: www.metacrawler.com
Key Search Word: atmosphere
Review: This is a terrific Web site. It is colorful, has great pictures, and is easy to use. It also provides many interesting and fun facts on a variety of subjects dealing with nature, energy, and technology. Across the top of each page are links to topics covered on the home page. They are Home, Nature, Energy and Technology, Articles, Projects, Goods, Media, Tours, E-cards, Ecoworld.org, About Ecoworld, Newsletters, and Register. Nature, for example, has links for air, water, earth, plants, trees, animals, and people. Each of these pages has a column on the left that will link you to all of the other subjects in the nature category. Also in this column are links that will give you data and images for your subject and a link where you can connect with articles and issues related to your subject. You can send e-cards as well as join ECOworld. In the right column are links to sites that sell merchandise related to the subject. The main body of each page gives informative and interesting facts on the subject. At the bottom of the page, there are links to the other pages with information on the subject you are viewing; just click on the page number to get there. I really liked this site. I thought it was easy and fun to use; anyone who is 9 or 10 years old should be able to use it with success. I would say that this site is fine for ages 10 to adult.

Title: The Weather Dude
URL: http://www.wxdude.com
Grade Level: Grade 3 and up
Best Search Engine: www.metacrawler.com
Key Search Words: atmosphere, elementary
Review: This was such a fun site! It's an educational site all about weather that's designed for kids and also, for their parents and teachers. It's "hosted" by Nick Walker, one of the meteorologists from *The Weather Channel* on TV. There are grants for teachers to win, songs for kids to download, and books for parents to order; there's even a current forecast for your local area when you input your zip code! Kid-friendly icons (like smiley faces and apples) all along the left of the page in-

struct you where to click for info on weather vocabulary words, how to become a meteorologist, weather records and averages, various weather topics and phenomena, and, to just ask a question of Nick, if you want. There are also links to other child-friendly weather Web sites at the bottom. I clicked on "Meteorology topics A–Z" and then on the sentence "A is for atmosphere; that's where it all begins." It told me about the six layers of the atmosphere, each with distinct characteristics, composition, and density. It let me click on each layer's icon to find out more; when I clicked on "troposphere," I learned that this is where weather occurs and where planes fly and that is the densest layer. It explained why temperature drops as your elevation increases, too. I personally learned a lot from this site that I'd forgotten about, and I honestly had fun doing it! It's easy enough for children to maneuver around as well and is packed with pictures and music to keep their attention. Visit this site; it's four stars!

Title: The Earth's Atmosphere
URL: http://www.windows.ucar.edu/tour/link=/earth/Atmosphere/ layers.html
Grade Level: Grade 3 to adults; great for teachers
Best Search Engine: www.metacrawler.com
Key Search Words: earth's atmosphere
Review: A bar at the top of most pages of this site offers three choices for different levels of users: Beginner, Intermediate, and Advanced. The point size of the text is larger and the language is simpler in Beginner. The Advanced version has a bit more information but is not too technical for students in middle school. The links to terms contained in the text (life, greenhouse effect, layers, weather) also offer the user the opportunity to switch among levels once the topic is reached. Some of the links (to the crossword puzzle and the chart of the solar system) allow the user to switch among levels once at the topic. The following are the links from "The Earth's Atmosphere" page: Weather Crossword Puzzle (your browser must be Java-enabled to run this game); Image showing the temperature of the atmosphere throughout different layers; Image showing the layers of the atmosphere with emphasis on the ionized layers; Introduction to the Atmosphere content; and Activities from Project LEARN. The following are the activities contained in this section: "Activity 1: The Goldilocks Principle—A Model of Atmospheric Gases"; "Activity 2: How High Does the Atmosphere Go?"; "Activity 3: It's Just a Phase—Water as a Solid, Liquid, and Gas"; "Activity 4: The Water Cycle"; "Activity 5: Atmospheric Processes—Radiation";

"Activity 6: Atmospheric Processes—Conduction"; "Activity 7: Atmospheric Processes—Convection Global Climate Change content and activities from Project LEARN." The following are activities contained in this section: "Activity 14: What Do Concentrations Mean? Comparing Concentrations of Gases in Our Atmosphere"; "Activity 15: What is the Carbon Cycle?"; "Activity 16: Dinosaur Breath"; "Activity 17: Where in the World is Carbon Dioxide?"; "Activity 18: Transpiration: How Much Water Does a Tree Transpire in One Day?"; "Activity 19: Wind Dynamics and Forests"; "Activity 20: Human Activity and Climate Change." "Introduction to the Atmosphere" and "Global Climate Change" are sections (like units containing an overview of concepts, text and graphics), and links to seven activities (shown above), which are full lesson plans for experiments including background, learning goals, alignment to national science standards, grade level, materials, procedure, assessment ideas, and modifications. If you link to "Sun" from the table, you will reach an entire section on "The Sun," as shown: "The Sun is the closest star to Earth and is the center of our solar system. A giant, spinning ball of very hot gas, the Sun is fueled by nuclear fusion reactions. The light from the Sun heats our world and makes life possible. The Sun is also an active star that displays sunspots, solar flares, erupting prominences, and coronal mass ejections. These phenomena impact our near-Earth space environment and determine our 'space weather.' " This is a very attractive full-color site, with color photographs and color graphics. The graphics add a great deal of information and interest to the text. The links available on "The Earth's Atmosphere" page alone lead to a wealth of related information; some of the destination pages have links to other topics that I didn't even explore, such as a unit on "Introduction to Ozone." The site is easy to navigate. The activities in the "Introduction to the Atmosphere" and "Global Climate Change" units are complete lesson plans, wonderful resources for teachers. What I especially like about this site is the opportunity to access the information at three levels of expertise and to readily switch among them if you want either a more robust or a simplified explanation of the topic. This makes the site appropriate for younger students as well as older ones. This is an impressive site in both scope and execution.

Title: Atmosphere
URL: atschool.eduweb.co.uk/kingworc/departments/geography/
     nottingham/atmosphere/pages/atmosphere.html
Grade Level: Middle school

Best Search Engine: www.yahoo.com

Key Search Words: atmosphere, earth science

Review: This is a cute Web site for students wanting to learn about the atmosphere. It was made by two students at the University of Nottingham as a project counting toward their degree in physics. The home page is easily navigated, as there are only six links to choose from. They are Key Stage 3, gcse, a-level, images, index, and help. The first three links are all related to questions having to do with the atmosphere. For example, if the student clicks on Key Stage 3 he will enter a page that tells him about atmospheric structure. There are definitions and graphs that explain all of the layers of the atmosphere in easy-to-understand terms. The graphics are useful to the students too because it helps them see the order to the layers of the atmosphere. The images link is really cool because it lets students see a satellite picture of the Earth and a picture of what a hurricane looks like from space. And the index link is really helpful because the authors of this Web site made an index, so a student can look alphabetically for the word he or she is researching and then click on the link next to it and get the information. This Web site would be great for sixth- to eighth-graders because it does not require a lot of clicking around to get the information they are looking for. And they would like looking at the graphics because it makes the subject so much easier to understand.

Title: Zoom Astronomy

URL: http://www.enchantedlearning.com/subjects/astronomy

Grade Level: Pre-K through Adult

Best Search Engine: www.google.com

Key Search Words: atmosphere and kids

Review: This site, sponsored by enchantedlearning.com, was great! I thought it was very interesting because it provides information about the atmosphere on all of the planets, not only the Earth's. The Zoom Astronomy home page contains an extensive amount of information about outer space. It is designed for people of all ages and levels of comprehension. It has an easy-to-use structure that allows readers to start at a basic level on each topic and then to progress to much more advanced information as desired, simply by clicking on links. The table of contents as listed on the page offers the Our Solar System, The Sun, The Planets, The Moon, Asteroids, Comets, and Stars. By clicking on any of those links, the user is taken to additional detailed pages about the subject. On each page you will find headings (links) that will direct the learner to even more information about the topic. These pages are well organized and all offer entertaining graphics. Learning activities can also be accessed through the links on the subject pages. Finding out about the Earth's atmosphere was simple. I clicked on "The Planets" link. Then chose the "Earth" link from the headings at the top of the page. Next, I selected the "Atmosphere" link from the Earth's page headings (1 of 15). Here, I found picture graphs and information about the Earth's layers and the atmosphere's formation, as well as a link to the NASA Web site on atmosphere. I thought this site was fantastic. Although a bit overwhelming, with the vast amounts of information, the site was well organized and very easy to navigate. The learning possibilities appear to be endless.

Title: The Atmosphere
URL: www.liftoff.msfc.nasa.gov/academy/space/atmophere.html
Grade Level: Grades 4–5
Best Search Engine: www.google.com
Key Search Word: atmosphere
Review: The site begins with a detailed description of the atmosphere. The text is straightforward and easy to follow. The site continues with an explanation of each layer of the atmosphere. At the bottom of the page there is a chart outlining the composition of the atmosphere. The last section briefly describes the Exosphere. The site is only one page, with no additional sites to click on; however, the concepts of the atmosphere are explained in a nontechnical, clear manner. I liked this site and learned a lot about the atmosphere.

Title: CIRA
URL: http://www.cira.colostate.edu
Grade Level: College and above
Best Search Engine: www.yahoo.com
Key Search Word: atmosphere
Review: This Web site is detailed. There are many links to click on to go to different pages. This Web site was generated at Colorado State University. On the left side of the pages are eight links: Introduction and History, Fellowships, CIRA Personnel, Infrastructure, Satellite Earth Station, Field Experiments, a newsletter, and employment opportunities. We learn from this Web site that CIRA means Cooperative Institute for Research in the Atmosphere. In the middle section of the page is a different title that you can click on to find information. The title that you click on has a picture that coincides with the name. To

the right of the names is a picture and information on the topic you choose. At the bottom of the Web site are other CIRA links. I really liked this Web site. There was much to look at and to stay interested with while reading. I also liked the setup of the Web site.

Blizzards

Title: Melissa's Webpage about Blizzards
URL: http://www.oars.utk.edu/volweb/Schools/sumnercs/ellism/mel.htm
Grade Level: Grades 1–8
Best Search Engine: www.google.com
Key Search Words: why do blizzards occur?
Review: This site was developed by a seventh-grade student named Melissa, in the Ellis School in Nashville, Tennessee, as part of a unit on severe weather. The site offers basic information about blizzards. The textual information and graphics provided offer understandable information about blizzards for elementary-age students. This page offers the following 10 links: What are blizzard conditions?, Advanced warnings, Why are blizzards hazardous?, How snow forms, Where in the world do blizzards happen?, Why do blizzards occur? The effects of winter weather, What do winter storms also bring?, Some of the biggest blizzards, and Return to weather homepage. Each link summarizes the topic listed for that link. For example, when you click on the link for "What are blizzard conditions?" a small bulleted description of blizzard conditions is given, followed by a listing of the other links provided on the home page. This structure allows for easy navigation. The link for "Return to weather homepage" describes other severe weather conditions also posted by students of the Ellis School. The graphics on the pages relate to the topic and in some instances help the reader better understand the material presented. This is a useful link for elementary students because it gives basic answers to explain what blizzards are, why they occur, what happens to people in blizzards, and blizzards that occurred in the past. It could also be useful for students studying a unit on weather because it not only provides information on blizzards, but also links students to other weather pages. The link to blizzards in the past may be useful to teachers interested in creating an interdisciplinary unit on weather and history. This link could also be used in elementary computer classes as an example of how to link science and technology.

Title: Blizzards and Snow Theme Page
URL: http://www.cln.org/themes/blizzards.html

Grade Level: Kindergarten to adult
Best Search Engine: www.metacrawler.com
Key Search Words: blizzards and education
Review: This site is a collection of lessons that will assist the study of blizzards and snow in your classroom. The Blizzard and Snow Theme Page is a link from the Community Learning Network, which is a site designed to help K–12 teachers integrate technology into the classroom. The Community Learning Network site has 5,800 annotated links to other educational sites with free resources and all organized by theme pages and keyword search. The Blizzards and Snow Theme Page has more than 30 links with lesson ideas for every grade level. For example, you can learn how to build an igloo or measure the density of water, ice, and snow. Also, this site has links to blizzard statistics from the past 50 years. This site is well organized by theme and grade level. A brief description under each link explains the objective and grade level. Overall, this Web site seems that it would be most valuable for educators looking to explore snow in their lessons on weather and Earth Science.

Title: Blizzards
URL: http: //www.weather.com/encyclopedia/winter/blizzard.html
Grade Level: Grade 6 and up
Best Search Engine: www.metacrawler.com
Key Search Word: blizzards
Review: The host site for this URL is the main weather.com page. This site seems to have everything you wanted to know about weather and then some. It does, however, take some time to find the topic you are looking for and to go through several pages going deeper into the site. For example, the entry screen from the Metacrawler Blizzard suggestions brings the viewer into the weather.com Safety Index. I wasn't looking for safety information so I went to the top of the page, where 17 categories are listed. I selected the Weather Encyclopedia, which takes you to its main page. There in the body of the page are another group of topics with drop-down menus. Look at the menu under Winter Storms and select Types of Winter storms. On the list at the bottom of the page select Blizzards; this will finally get you to the correct information page (and to the URL listed above). This is a really good site for information but it does take a little patience and some thought to get around. If you have a slow computer it will take a lot of patience to load each page as you navigate around looking for your topic.

Title: Blizzards
URL: http://teacher.scholastic.com/activities/wwatch
Grade Level: Grade 1 and up
Best Search Engine: www.google.com
Key Search Words: blizzards and education
Review: The front page is Weather Watch/Blizzards. The top part of the
screen gives you three links: All About Hurricanes, Ask our Weather
Expert, and Hurricane News. Clicking on the first link, All about Hur-
ricanes, gives you one and a half pages of facts and descriptions of a hurricane. Located on the left side of that screen are four bullets: Track a
Hurricane, Hurricane Trivia, Internet Field Trip Pr K–3, and Internet
Field Trip 4–8. These bullets were great. They gave you plenty of Science and Art Projects. The second link, Ask our Weather expert, gives
you the opportunity to ask questions to real weather experts. This
would be great for young students. The third link, Hurricane News,
gives you interesting real-life hurricane stories. Back on the opening
page, in the Interest Section, there are two links. The first link, Extreme Weather Rescue Starter, helps you learn about ways to become
prepared for a hurricane. The second link is Scholastic Weather Reporters. This section brings you seasonal classroom experiments and
hands-on activities. It is great for kids of all ages. It teaches kids about
the various weather forecasting tools, and you can share your experiment and analyze your experiments. This site was interesting and informative. It gives teachers and students fun projects.

Title: The Cryosphere: Where the World Is Frozen
URL: http://nsidc.org/cryosphere/index.html
Grade Level: Grade 6 through college
Best Search Engine: www.yahoo.com
Key Search Words: blizzards earth science
Review: This is a well-organized site with an abundance of good and in-
teresting info. How to navigate: Once you are on the homepage,
there are quick links in the box on the right that take you to infor-
mation concerning Arctic Climatology and Meteorology, Glaciers,
Ice Shelves and Icebergs, Avalanches and Blizzards, and The State of
the Cryosphere.

Climate

Title: Geographic.org
URL: www.geographic.org

Grade Level: Grade school and junior high
Best Search Engine: www.metacrawler.com
Key Search Words: climate, elementary
Review: This Web site was designed for elementary through junior high
    school students, their parents, and their teachers. It gives information
    on geography, weather, different countries, maps, and so on. I clicked
    on the subject "geography" on the main page and then on the word
    "climate" at the next page, but I later realized that I could have also
    just typed the word "climate" into the "search" box on the home page
    and gotten the same results. It gave me pictures, a glossary of words
    relating to the term "climate" and a chance to find out more, which I
    did by clicking on that icon. It then told me how to make simple
    weather instruments such as pinwheels and wind chimes, and it gave
    me tips on easy activities to do with children, to teach weather, like
    just going outside and watching the clouds. It was easy for me to nav-
    igate the site, but other than the glossary, I think it's most appropriate
    for adults who want teaching ideas, rather than kids looking for fun
    stuff to do.

Title: Climate Diagnostics Center
URL: http://www.cdc.noaa.gov
Grade Level: High school and college
Best Search Engine: www.dogpile.com
Key Search Words: the climate
Review: This Web site is created by the Climate Diagnostics Center
    (CDC). The CDC is responsible for understanding and predicting
    climate variability. The Web site gives details about the center itself
    and then supplies the reader with information about the general cli-
    mate. It provides a map allowing the reader to observe the climate in
    any part of the world. The most recent research on climate is pro-
    vided within the Web site and available to the reader. To the left of
    the page, the site provides the latest news and features that deal with
    the climate. The climate research spotlight is available with a simple
    click. Job openings all over the world, generally in the United States,
    pertaining to climate, are available to the reader. Seminars relating to
    climate that take place all over the world are listed with their loca-
    tion and topic. In addition, this Web site has a site index and search
    engine, making things easy to find. All in all, this site is an excellent
    source for high school and college students to use when information
    about the climate is needed.

Title: Program for Climate Model Diagnosis and Intercomparison
URL: http://www-pcmdi.llnl.gov
Grade Level: College and above
Best Search Engine: www.yahoo.com
Key Search Word: climate
Review: This is a well-designed Web site. It tells the readers all the infor-
    mation they need for finding the section they are looking for in the
    site and sends the reader there by different links. At the top of the
    site is a new area that has recently been adapted, and that is an-
    nouncements for the site. Under that section is a review of how the
    program the site talks about was developed. Following that history are
    different facts from the program. Underneath the facts and history are
    links to six projects. On the right side of the pages are different links
    about PCMDI, its location, staff, and such technical material as dy-
    namical core, model features, and model data. I like how this site
    gives both related sites and a search. I believe that this is important
    because you can look for all the information you want just by this
    Web site.

Title: Learn: Atmosphere Science Explorers
URL: http:/www.ucar.edu/learn/
Grade Level: Middle school science teachers; portions are good for stu-
    dents from middle school to adult
Best Search Engine: www.metacrawler.com
Key Search Words: weather AND climate
Review: This Web site is the Project Learn on-line teaching module
    for middle school science teachers and consists of seven sections:
    Introduction to the Atmosphere, Introduction to Climate, The
    "Greenhouse Effect, Global Climate Change, Introduction to Ozone,
    Stratospheric Ozone, and Tropospheric Ozone. The second, third, and
    fourth sections, taken together, provide comprehensive coverage of
    the topic "climate." Each section contains a general concepts over-
    view, introduction, concluding thoughts, and links to several activi-
    ties. The activities are full lesson plans including background, learning
    goals, alignment to national science standards, grade level(s), materi-
    als, procedure, assessment ideas, and modifications. The Introduction
    to Climate section includes these topics: How does climate differ from
    weather?, Climate Variability, and Paleoclimates, plus links to the
    following activities: "Activity 8: Differences Between Climate and
    Weather"; "Activity 9: Climate Variability"; "Activity 10: Paleo-

climates and Pollen"; and "Activity 11: Time and Cycles: Dendrochronology." The Greenhouse Effect section includes Solar Radiation, Greenhouse Gases, Greenhouse Effect, and links to the following activities: "Activity 12: What is a Greenhouse?" and "Activity 13: What Factors Impact a Greenhouse?" The Global Climate Change section includes Climates Past, Present Climates and Human Activity, Future Climates—The Great Uncertainty, and links to the following activities: "Activity 14: What Do Concentrations Mean? Comparing Concentrations of Gases in Our Atmosphere"; "Activity 15: What is the Carbon Cycle?"; "Activity 16: Dinosaur Breath"; "Activity 17: Where in the World is Carbon Dioxide?"; "Activity 18: Transpiration: How Much Water Does a Tree Transpire in One Day?"; "Activity 19: Wind Dynamics and Forests"; and "Activity 20: Human Activity and Climate Change." Each section contains color graphics, some of which are dynamic. This site is graphically appealing and easy to navigate. The only links are to the lesson plans. Although designed for teachers, the language is appropriate for students from middle school up. The topics are covered quite comprehensively via text and supporting graphics. This site is a good resource for someone who wants to understand climate, and it is a great resource for teachers.

Title: World Climate—Weather Rainfall and Temperature Date
URL: www.worldclimate.com/worldclimate/index.htm
Grade Level: Grade 8 to high school
Best Search Engine: www.metacrawler.com
Key Search Word: climate
Review: WorldClimate.com was created by a computer consultant. The home page allows you to search for the current climate in any town or city in the world. Just type in the name (in any language) and click on search. In the text, click on About World Climate to get more information on questions such as, what is world climate and how is the data compiled. The Web site is designed for general audiences, schools, and travelers who are interested in general weather patterns around the world. The data is provided without any warranty, and there is a disclaimer saying this site is not "suitable for professional or research use." The data provides historical averages, the typical weather in a month that is averaged over several years. There is a section on Sources; click and there is information on how and where the data on weather was compiled. At the bottom of the second page, click on Frequently Asked Questions, and there is a link to *The Old Farmer's Almanac*, where you can select any date from 1994 to pres-

ent in any area of the world for climate information. There is a map of the United States; can click on a state and there is information on weather stations. This site is historical in nature; it is not a forecasting tool. There is a section on definitions, such as Air Temperature, Average Maximum Temperature, and so forth. This site was interesting, but it is under development and more data and links will be added. I would recommend this site for eighth grade to high school students studying science or to provide information for a creative writing assignment.

Title: CMDL—Climate Monitoring and Diagnostics Laboratory
URL: http://www.cmdl.noaa.gov
Grade Level: Grade 8 and up
Best Search Engine: www.google.com
Key Search Word: climate
Review: This site, hosted by the U.S. Department of Commerce, is jam-packed with information and interesting subpages. The site's home page immediately pulls the visitor in by asking some important questions regarding our climate, the ozone layer, and air quality. Each question has text links to additional pages that go further in describing the keywords in the text that relate to the overall topic. As the user scrolls down, there are images of the Earth that represent the climate, ozone, and air quality; these lead to the same pages as the text links that provide definitions, charts, and other pertinent information. Farther down the page there are text links to current news about the program and its research. There are also a number of text links listed to sites that feature information on other topics of the group's research, such as aerosols and radiation, carbon cycle greenhouse effects, and more. These links are located just above a bar located across the bottom of the page that provides text links for easily navigating to other main pages/areas of the site. The same bar runs across the top portion of the page as well as all of the subpages. Down the entire left side of the page, there are pictures of the various observatories used by the research groups. When a picture is clicked on, you are taken to a page that shows a larger picture of the observatory as well as a description of the observatory and what research is being conducted there. Along the left of these secondary pages are a number of additional icons that lead to interesting pages that provide additional and more substantial information about the site, what life is like there, and the site's history. All of these pages are full of interesting information, great pictures of these amazing places, charts,

graphs, and other data relevant to climate studies vital to these areas and our world. This is a fantastic site. The pages highlighting the study/observatory sites are excellent. Each provides the user with great pictures, background information, research information, and much more to the point that one feels as if they are there, taking a tour of the site. This feeling is enhanced by the live cameras that can be accessed on each of the individual site pages. This comprehensive site could really be interesting for upper middle school students and above to dig deep into this topic and learn about what studies are being done throughout the world to monitor and protect our climate. To make using this site worthwhile, the teacher would want to allow a lot of time for students to explore the site—it really is a neat one.

Title: National Weather Service Prediction Service
URL: http://www.cpc.ncep.noaa.gov
Grade Level: High school
Best Search Engine: www.yahoo.com
Key Search Word: climate
Review: This Web site looks at many aspects of the weather and climate of the United States and other countries. When you first get to the home page there is a search tool. There was also a Top News of the Day section. I found this to be interesting and full of information. I began navigating through the Web site by clicking on Climate Highlights, which had many subheadings to choose from. I then went to U.S. Drought Assessment, which was full of information. This site was full of pictures, graphs and was inviting to read. I found the information and graphs to be right at the high school level. This site also had a section that would connect you to other links that were related to this topic. At the bottom of this home page there was a section providing links so that people can contact the National Weather Service or find out what the NWS is all about.

Title: United States Interactive Climate Pages
URL: http://www.cdc.noaa.gov/USclimate
Grade Level: High school
Best Search Engine: www.excite.com
Key Search Word: climate
Review: This page, created by NOAA-CIRES Climate Diagnostics Center, is devoted to predicting weather and graphing and plotting weather patterns. When you first enter the site there are three sections to click on. The first section is climate information for individ-

ual cities in the United States. When you get in the section, there is a map of the United States and you can click on any state. Then when the state map comes up you can click on a specific city and get climate information on that city. The second section is called climate maps for the United States. It lets you create a map of monthly U.S. temperatures and precipitation from 1895 to the present. It is really cool because the students can put in any year and get the weather charted out for them. It is easy to fill in the boxes, and it is not confusing at all. The third section is more of a reference section because it lets you find definitions to weather terms and other Web sites that contain climate and weather data. This is a good site to use for secondary students and above. It is really hands-on because it teaches them how to plot weather trends and it offers many examples of past plotted weather patterns. I would not recommend this site to elementary school students because it uses many technological terms that they might not understand.

Title: World Climate
URL: http://www.worldclimate.com/climate/index.htm
Grade Level: Middle school
Best Search Engine: www.metacrawler.com
Key Search Word: climate
Review: I thought this site was particularly interesting in that it provides detailed climate information about any place in the world. Upon entering a city/town in the required field, the user will be taken to a page with a series of links. These links, organized by the nearest surrounding areas, outline the various aspects of the climate for the area. I entered Long Branch and was taken to a page broken into various towns in New Jersey. I scrolled down to find the Long Branch section and links to the following information was offered: Average Max Temp, Average Temp, Average Min Temp, Heating and Cooling Degree Days, and Average Rainfall. When clicking on one of these links you will be taken to an information page. For example, when I clicked on Average Rainfall I was taken to a chart of the rainfall in Long Branch over the course of this past year. Here you can also find a link to the current weather for the area. I thought this site was interesting, especially to use as interactive media in the classroom. Although there are no exciting graphics on the site, the information can make studying far away places more relevant for the students. The Web site is easy to navigate and the information is direct and factual. This is a great resource for a quick, up-to-date climate reference.

Title: Global Warming/Climate Change Theme Page
URL: http://www.cln.org/themes/global_warming.html
Grade Level: High school through adult
Best Search Engine: www.google.com
Key Search Words: earth sciences, climate
Review: Once I found the Global Warming/Climate Change Theme
   Page, I scrolled down, reviewing the sites for one that I might be in-
   terested in until I found United Nations Framework Convention on
   Climate Change. Listings under this site are Beginner's Guide to
   the Convention, A description of environmental problems that the
   world is facing and what we are doing about it through the Convention, Climate Change Information Kit Thirty fact sheets (also in
   downloadable format), which cover: Understanding of the Climate
   System; Facing the Consequences; The Climate Change Convention; Limiting Greenhouse Gas Emissions; and Facts and Figures,
   [The] Convention and the Kyoto Protocol Full text of the protocol
   along with information on the status of its implementation, and
   Country Information Snapshots of actions taken by a large number
   of countries to meet their domestic and international climate
   change commitments. For the information I was interested in finding, I chose to look under Climate Change Information Kit. This
   link has several topics from which to choose. It begins with a Fore-
   word and has such topics as Understanding the climate system, Facing the consequences, The Climate Change Convention, Limiting
   greenhouse gas emissions, and Facts and figures. Each page gives detailed information and facts on each of the subjects. This is definitely material written for adults. It is loaded with facts and highly
   informative. This may be a lot for even high school students at
   lower levels.

Clouds

Title: PSC Meteorology Program Cloud Boutique
URL: http://vortex.plymouth.edu/clouds.html
Grade Level: Grade 3 through college
Best Search Engine: www.yahoo.com
Key Search Words: clouds earth science
Review: Great site for basic info on clouds and quick links to beautiful,
   large pictures. Once you are on the home page, scroll down and it
   gives you an outline for how the site is set up. To get the pictures, just
   click on the links.

Title: Clouds
URL: http://seaborg.nmu.edu/clouds
Grade Level: Grade 2 and up
Best Search Engine: www.excite.com
Key Search Word: clouds
Review: I stumbled onto this wonderful site the first try. This Web site appears to be just an informative site uploaded by someone who sees the educational value in clouds. Carl Wozniak, the site's author, did a wonderful job putting this site together. As we look at the home page, we're presented with six options to go to. They are Initial Cloud Formation, Cloud Types, Cloud Pictures, Other Phenomena, Glossary, and Links to Other Weather and Earth Science sites. After choosing the Initial Cloud Formation link, we are presented with an exception overview of how clouds form, including discussions on "Why is the Sky Blue?" and "And the Clouds are White Because . . . ?." After leaving this area, go to the link for Cloud Types. The site explains how clouds are named for their general appearance and height in the atmosphere. The page then lists nine types of clouds, their descriptions, and their height in the air. The next part of the Web site was my favorite: Cloud Pictures. I'm not sure if Carl Wozniak is a professional photographer in his spare time, but he should be. The site gives a great number of beautiful, high-resolution images. Each photograph is labeled and described. This site is worth taking a look at just to see these pictures. The Other Phenomena link didn't work for me. I assume the site must have not been updated lately and perhaps is just an old link. If at any point throughout your navigation through this site, you don't understand a word, you can go to the glossary and every term is laid out for you there. Some great Earth Science links are listed under the links section and are worth taking a look at as well. Overall, this is a great site for children to learn about clouds. It's also a great site for the unenlightened adult to see big, bright, beautiful photographs and to further their knowledge of clouds.

Title: Classifying Clouds
URL: http://weather.about.com/library/weekly/aa032802ahtm?iam=
    metacrawl_1&terms=%2Bclouds
Grade Level: Fifth to adult
Best Search Engine: www.metacrawler.com
Key Search Word: clouds
Review: This Web site is sponsored by www.about.com and can be useful for anyone who is interested in studying the weather. Along the left

side of the site is an index of all of the essentials of weather. For example, there is a link to basic meteorology, weather forecasts, atmospheres, blizzards, climate, precipitation, air pollution, and many more. Once you click on one of the links, the right side of the site will provide you with more information than you could ever imagine about that topic. When I clicked on clouds, the site listed all of the different cloud types with a description about each type. Also, it provided me with a quick reference guide to understanding cloud formations and the different cloud types. The site had a great link to photographs of all of the cloud types, too. There were many Internet sites that I found for clouds, but this one seemed to be the easiest to navigate through and it provided the best pictures with the cloud descriptions.

Title: Clouds for Kids
URL: www.public.iastate.edu/~jswift/clouds.html
Grade Level: Grades 1–5
Best Search Engine: www.msn.com
Key Search Words: clouds, children
Review: The purpose of this Web site is to introduce young children to pictures and facts about the four types of clouds. It was designed by a college student named Jami Swift from Iowa State University and created in the hopes of assisting children at the elementary level. The home page includes a picture of each of the four types of clouds (cirrus, stratus, cumulus, and cumulonimbus) so that when a child clicks on the picture of their choice, a narrative description appears along with the scientific name. There is a link at the bottom students can also click on that prompts them to the question "What kind of Cloud do you see today?" Here students are encouraged to answer the question giving a brief description of what they see. Overall, I feel that this site is a great tool to assist in teaching a beginning lesson on clouds. The site is user friendly and can be mastered by younger children with little computer literacy.

Title: Clouds
URL: http://www.doc.mmu.ac.uk/aric/eae
Grade Level: Grade 6 and up
Best Search Engine: www.metacrawler.com
Key Search Word: clouds
Review: The master site here is the Encyclopedia of the Atmospheric Environment. This site is a compendium of topics on the atmosphere

and is a good resource for all things atmospheric. On the Entry page, select the "English" tab (if you want the site in English, or you can select French if you prefer). This takes you to the main page of the site. The left column lists a lot of atmospheric topics; select "weather" by double-clicking on the word. The new page has an extensive list of weather and meteorological topics down the right side (the original list on the left is still there for a quick transition to another topic). Each topic has at least one character next to it from *The Simpsons* TV program. The sites with two pictures (with the second one of Bart) tells the viewer that this is an alternative, less technical site written in simple terms for children. Once you finally reach the "Clouds" page there is a good, concise description of the types of clouds, and at the bottom of the page there are links to photos of each type of cloud discussed. In the text of the site each type of cloud is highlighted and underlined. By clicking on the word for specific type of cloud the viewer is taken to a new page with a more detailed discussion. This is a good site that accommodates a wide variety of viewers without "dumbing down" the information. Anything related to weather can be located on the master site with ease.

Title: Clouds
URL: http://www.pals.iastate.edu/carlson
Grade Level: All
Best Search Engine: www.google.com
Key Search Words: clouds and kids
Review: The home page gives a brief introduction about the creator of the Web site and a picture of him pointing at a beautiful cluster of clouds. When you scroll down to the bottom of the screen there is a Start button. Double-click on the button and it takes you to a page of a cloud that he calls the "Energizer Rabbit." What a neat picture! Located on the right hand side of the screen is a Text button. There is a brief description of the cloud. On the right hand side of the button is a Next button. Double-click on the button and it gives you eight categories of photos. 1. Clouds that look like things: located in a column are these subcategories: Animals, birds, insects, sequences, people, prehistoric, and scary. All of these selections bring you to amazing real photographs. 2. Unusual Clouds. 3. Pretty Clouds. 4. Sunsets. 5. Ice. 6. Clouds That I Can't Figure Out, Can You? 7. Spring Time Scenes of Iowa State University. 8. Contributed Photos. Located on the bottom of the screen is a Links button. When you double-click on the button it takes you to three Web site choices: Clouds Gazing-Art

in the Sky, Weather Works, and In the Clouds. This Web site was really beautiful. It offers hundreds of photographs of clouds.

Title: Boat Safe Kids—How to Be a Storm Spotter
URL: http://www.boatsafe.com/kids/weather1.htm
Grade Level: Kids of all ages (this includes grown ups)
Best Search Engine: www.google.com
Key Search Words: clouds and kids
Review: This site is a subpage of the Boat Safe Kids main page. This Web
    site is dedicated solely to information about sailing, specifically for
    children. This particular page is named "How to Be a Storm Spotter."
    It gives readers a brief paragraph about why weather predictions are
    helpful to sailors. Then it tells how you can spot a storm by looking at
    the clouds. In short, by knowing what types of clouds are in the air,
    you can spot a storm coming and get to safety. The page then shows
    beautiful pictures with different classifications of clouds and offers
    mnemonic devices to remember the cloud names. The textual infor-
    mation is concise and easy for children to understand. Important
    glossary terms are bolded so they can easily be seen. At the bottom of
    the page, tips that a storm is coming are given with a graphic, so kids
    can see what to look for. Safety tips for what to do in a storm are also
    given. A link back to the home page is given at the end of the lesson.
    The best parts of this page are the graphics. Each of the 13 types of
    clouds discussed has a gorgeous picture of that cloud next to it. A pic-
    ture really is worth a thousand words on this page. This is an excel-
    lent page for anyone interested in gaining information on clouds or
    anyone who just wants to appreciate the beauty nature provides.

Coastal Processes

Title: Science Junction
URL: http://www.ncsu.edu/coast/educator/index/html
Grade Level: Elementary to high school
Best Search Engine: www.metacrawler.com
Key Search Words: coastal processes, elementary
Review: This Web site explores coastal processes and was created based on
    the goals stated in the national science education standards. It can be
    used by teachers and children in environmental science classes of ele-
    mentary, middle, and high school levels. It integrates social studies, ge-
    ography, language arts, and media as well. Various topics and teaching
    suggestions are given once you log on to the main page. I scrolled down

to the section on primary school science curriculum, kindergarten–fifth grades. Once here, I was able to see pictures of different panoramas, where I could see similarities and differences in the amounts of sand, waves, and plant covers between the oceansides and marshsides of barrier islands. I also learned about inlets, coastal rivers and sounds. If I were teaching children, I could definitely use this. I would have them do something as simple as drawing their own barrier island, or if they were older, I could have them do their own in-depth research on beach erosion or sewage disposal, where they might put themselves into a coastal oceanographer's shoes. There's a lot to do on this site; definitely worthwhile for classes and families to visit.

Title: USGS—Science for a Changing World—Coastal Change
URL: http://pubs.usgs.gov/circular/c1075/change.html
Grade Level: 6th grade and up
Best Search Engine: www.metacrawler.com
Key Search Words: coastal processes education
Review: This page, hosted by the U.S. Geological Survey of the U.S. Department of the Interior, provides a comprehensive explanation of coastal processes and change. The page is somewhat conservative in its overall design and although it mostly consists of text, there are several pictures included that allow the visitors to see examples of some of the information being described. The page begins with a thorough description of what coastal change is as well as its effects. Immediately below this opening paragraph is the first of several text links that take the user to a subpage that further details an example of the topic being discussed. These subpages, throughout this site, provide interesting examples of the coastal processes in action and its effects on these areas. The page continues to scroll down and explore the variety of forces and factors that play a role in the coastal process, such as waves, tides, weather, water level changes, coastal vegetation, and more. Each of these is discussed in detail, and examples of historical events or records reinforce the information. The coastal process is then reviewed in regards to specific and unique areas such as mainland beaches, wetlands, barrier beaches, and lakeshores. The page then goes on to examine the effects of human intervention on this natural process. Each topic has a great deal of information that assists in better understanding the entire process and its effects. This page is extremely informative and would be a great resource for students and teachers seeking information on coastal processes. The information is easy to understand and follow as well as very well presented. The text links drill deeper into the topic and en-

hance the page's usefulness in thoroughly explaining coastal processes. I think this is a great page to reference when seeking information on coastal processes because it is loaded with great information, photos, diagrams, and concrete examples related to the topic.

Title: Coastal Processes
URL: http://www.geography-games.com/coastal_skills.htm
Grade Level: Proficient readers, grade 4 and up
Best Search Engine: http://www.metacrawler.com
Key Search Words: coastal processes
Review: This particular topic is identified as "Coastal Skills" on the menu at the left of the screen. Across the top of the screen is a display that reads from "Worksheet" at the left, then Page 1 through Page 10 at the far right. The following lists the topics by page number. Worksheet: A file to download which is a fill-in-the-blank summary of each topic, with links to additional Web sites; an animation of stack creation and sites to be used for an extension activity to write a report on the problems caused by coastal erosion to coastal communities. 1: Introduction (graphic and 11 lines of text). 2: The Coast (graphic and 6 lines of text). 3: Parts of the Coast (graphic and 7 lines of text). 4: Cave, Arch, and Stack (graphic and 12 lines of text). 5: Wave Cut Platform (graphic and 7 lines of text). 6. Blow Holes (graphic and 7 lines of text). 7: Parent Rock Structure (graphic and 12 lines of text). 8: Wave Measurement (graphic and 5 lines of text). 9: Wave Types (graphic and 9 lines of text). 10: Longshore Drift (graphic and 12 lines of text). Each page contains a black-and-white line drawing illustrating the topic in addition to the text. There are no links in this site other than those on the Worksheet. Although this is not the most graphically exciting site, it presents the topic of coastal processes concisely in clear language with effective black-and-white graphics. It is easy to navigate from page to page. The site contains all of the data without having to use links to reference other pieces of information, which is a less complicated presentation for younger students. This is a good basic introduction to the topic, whether it is used by an adult or by a middle school student.

Title: Coast—A Resource Guide for Oceanography and Coastal Processes
URL: http://www.coast-nopp.org/resource_guide/index.html
Grade Level: Teachers teaching K–12
Best Search Engine: www.yahoo.com

Key Search Words: coastal processes and elementary education

Review: This is a great site for teachers if they are interested in doing lessons on coastal processes and oceanography. It was created by the U.S. Navy, the National Oceanic and Atmospheric Administration, the Department of the Interior, and the National Marine Educators Association. It was created to provide an interactive, comprehensive approach to teaching coastal processes. The site is easily navigated. The teacher can choose right away if she wants to use the elementary and middle school lessons, or the high school lessons. If the teacher clicks on the elementary lessons she will get to choose from six different units: Physical Parameters, Plate Tectonics, Marine and Aquatic Habitats, Marine and Aquatic Pollution, Marine and Aquatic Resources, and Deep Sea Technology. Then each unit is broken down into about 15 different lessons, that have activities, objectives, and evaluations all written down for you. The high school section has units on: Classification Systems, Evolutionary Adaptations, Marine Communities, Biotic Human Interactions, and Hydraulic Systems to name a few. They are set up the same way as the elementary units. The site also has a section for site usage tips and a Web links database. The thing I like about this site, as a future teacher, is the way it is organized. When you click on a link it takes you right to the lesson; you don't have to click around 500 times to get what you want. I would definitely recommend this site to any science teacher out there because the organizations involved have put a lot of work into developing these lessons and it shows.

Title: Related Oceanographic Information—Coastal Processes
URL: http://bigfoot.wes.army.mil/c112.html
Grade Level: Middle School
Best Search Engine: www.metacrawler.com
Key Search Words: coastal processes and kids
Review: I thought that this site was interesting because it offers quite a few links to various oceanography topics. On the home page, the following links, as well as many more, are listed; Clean Ocean Action, Coastal Briefs, Coastal Change, Coastal Zone Management Program, Land-Ocean Interaction in the Coastal Zone, and Net Coast. When I clicked on the Clean Ocean Action link I was taken to the site about the water quality of the marine waters of New York and New Jersey. Here, I clicked on the "Dive In" link and was taken to a page with a large menu. The menu includes links such as: Action Alerts, Resources, Beach Sweeps, Fun Stuff, and Project B.E.A.C.H. I

thought this was great because it provides information that is directly affecting the local waters of New York and New Jersey. In the Fun Stuff section the user can find games and activities such as Fish Silhouettes and Mystery Marine Animals. Although the main page of this site simply directs the user to many other sites the information is very extensive. The sites are well organized and are easy to navigate. I recommend http://bigfoot.wes.army.mil/c112.html to anyone, middle school and up!

Title: Shorelines and Coastal Processes
URL: http://www.dc.peachnet.edu/~pgore/geology/geo101/coastal.htm
Grade Level: Middle school to high school
Best Search Engine: www.dogpile.com
Key Search Words: coastal processes
Review: This Web site was created by Pamela Gore from Georgia Perimeter College. Water movements such as waves, tides and currents, and sea level changes are some of the topics covered within the site relating to coastal processes. To help illustrate and exemplify the information being discussed, detailed pictures and illustrations are provided to the reader. Words pertaining to coastal processes such as *wave height*, *wave base*, *wave refraction*, and *water particle motion* are defined in great detail to help the reader better understand the information being discussed. One great but different aspect of this site is that the author provides the reader with a list of movies that the reader can watch to better understand the information provided within the Web site. The author describes in great detail at what point in the movie one should pay close attention, for this is what is exemplifying the information being discussed. In addition to movies, various types of additional resources are provided for the reader to locate helping them to better educate their knowledge on coastal processes. This Web site does an excellent job at catching the eye of the reader for every topic. All in all, this site is an excellent resource for students to use to better educate themselves on the topic of coastal processes.

Title: Coastal Processes and Landforms
URL: http://www.geog.buffalo.edu
Grade Level: College
Best Search Engine: www.google.com
Key Search Words: elementary science, coastal processes
Review: This is without a doubt intended for a college student or adult. It is full of informative text with few illustrations. It discusses in detail

all of the coastal processes that operate on the coastline. The page begins by defining the terms *shoreline* and *coastline* and then goes on to discuss the coastal zone. The next section of text discusses the topic of the forces that shape coasts. The next topic on the page is coastal erosion and sediment transport and deposition. The next subject area discussed is coastal landforms, followed by the last topic: types of coastlines. Each topic gives detailed information about that subject and all relevant definitions needed to understand the process. If I were looking to study the coast and how it forms and changes as well as develop an understanding of the processes which take place, many of my questions would be answered by this site.

Cosmology

Title: Comets
URL: http://csep10.phys.utk.edu/astr161/Lect/comets/comets.html
Grade Level: High school through college
Best Search Engine: www.metacrawler.com
Key Search Words: comets + science + education
Review: There are quite a few very fine astronomy Web sites, in fact a veritable glut in comparison to a subject such as vectors. Some are tutorial based such as Gene Smith's Astronomy Tutorial (located at the University of California, San Diego center for astrophysics and space sciences, http://casswww.ucsd.edu/public/tutorial/BB.html). Other excellent astronomical Internet sites have been developed as a result of college or university courses and so it is with Astronomy 161—the Solar System, which is a 22-segment course that begins with an overview of the sky and planets, then proceeds to the History of Astronomy from Ancient Times to Einstein's Theory of Relativity, then turns the survey from the celestial coordinate system, the eclipses, and a examination of the solar systems planets and Planetesmals (including comets, asteroids, and meteors). The section on comets can be found at http://csep10.phys.utk.edu/astr161/Lect/comets/comets. html. Crisp color photographs, which print out slide-quality prints (at least from my printer), can be easily incorporated into a basic astronomical library. The segment includes linked Web pages on the nature and observation of comets (including cometary orbits, and "plotting your own orbits" (using solar system live software), the coma and tail. Here, the nucleus of Comet Halley is portrayed (it was actually photographed by a spacecraft). In Comet Hyakutake there are links to the Hubble Space Telescope (HST), observations of

Hyakutake and even movies (AVI and QuickTime format) of Hyakutake. There is even an extended section on my favorite, most memorable comet, Hale-Bopp, which I saw on a crisp, cold morning from a lightless campus thoroughfare at Montana State University (there was an electrical renovation taking place during a student vacation). I recall the comet was so clear and bright and I could even make out a tail. There is a section on Halley's Comet, which was a real visual bust as far as I could tell in 1986. Web site views show Halley's Orbit in 2024, when it will be at its aphelion (greatest distance from the sun) and in another view, how elliptical Halley's Orbit truly is. There is also a very nice addition on the 1908 Tunguska Event in Siberia and Comet Shoemaker-Levy 9's encounter with Jupiter. The segment ties up with a page on "Small water comets." By far, the site is well organized and much information is available at the click of a mouse, with navigation between pages and lessons user friendly.

Title: Cosmology
URL: http://casswww.ucsd.edu/public/astroed
Grade Level: High school to college
Best Search Engine: www.metacrawler.com
Key Search Words: cosmology + science + education
Review: This review is dedicated to Gene Smith's Astronomy Tutorial (GSAT), which is one of nine major headings on the site map of the astronomy Web site at the University of California. Besides GSAT, there are many other places to explore onsite including "Images & animations," "Astronomy Websites," "Astronomy courses on the web," "Astronomy Curriculum Resources," "Text Documents," "Astronomy Publications & Organizations," and "related educational sites." Additionally, the GSAT site covers over 20 separate topics on space and astronomical related subjects. Just look at the list (this is not exhaustive: Refractors; Reflectors; Stellar Spectra; Stellar Evolution; General Relativity; Galaxies; Atomic Structure; Supernovae, Black holes; Dark Matter; Quasars; The Big Bang; And . . . Cosmology . . . this review's feature subject! This section of GSAT opens with "cosmology: The structure & future of the universe and a few well posed questions that Wayne and Garth might ask at a philosophical moment: (1) If the Universe is expanding, what is it expanding into? (2) What happened before the Big Bang? (3) If everything is moving away from us, doesn't that put us at the center of the universe?" In succeeding sections, the Big Bang is examined (including nuclear reactions in the Big Bang), while a useful time-

line of events in the Big Bang and evidence for the Big Bang (e.g., cosmic background radiation) is provided for students and teachers. I browsed the site on this for over several hours (I always had an interest in astronomy since I was young) and still found there was a great deal I had to forgo looking at. What was so marvelous about this site is that it takes a middle ground between being an astronomical online encyclopedia and a teaching resource (as a tutorial) that could be used in class or perhaps even after school in a science club. Even if one had no access to a paper-based encyclopedia, one could still have access to an extensive, well thought out, astronomical resource on the Internet that emulates the breadth and depth of our solar system. Although one may be tempted to take such Internet sites for granted, don't. Even the limited Web-site building I've done tells me that a great deal of care and effort went into making this one. While you may have been at GSAT before, like I have, I still found browsing it, enjoyable and in places educational, even though I have a considerable knowledge base on the subject to begin with. Because of its depth, it also has room for multiple uses as a resource and extension for bringing the subject alive for students. While it is not the end-all of astronomical sites on the Web, it is an excellent place to start or continue one's search and investigation of the cosmos, our solar system and our place in life, the universe and everything (to borrow from Douglas Adams).

Title: Origins
URL: http://origins.jpl.nasa.gov
Grade Level: Grade 6 through college
Best Search Engine: www.yahoo.com
Key Search Words: cosmology earth science
Review: This is a wonderful site. There is a lot of profound knowledge broken down into easily understood pieces. This is NASA's attempt to solve two questions: "Where did we come from?" and "Are we alone?" How to navigate: on the homepage, there is a menu bar on the left of the screen. The site is also searchable, at the top right of the home page.

Title: Cosmology
URL: http://cybersleuth-kids.com/sleuth/Science/Space
Grade Level: K–12
Best Search Engine: www.google.com
Key Search Words: cosmology and kids

Review: This site is an amazing search engine for kids and teachers. On the home page the title is located on the top: "CyberSleuth Kids search the Internet. A Internet Search Guide for the K-12 student." Under the title are different options: Home, Fun and games, Science, Math, References, Art, Geography, and History. In the upper center of the Home page is a search button. This gives you the option to connect to any area of interest. When I typed in "cosmology," it gave me 75 sites of interest. Back to the home page; Located on the left hand of the screen is Classroom Clipart. Double-click on the button and it gives you a wide, cool selection of clip art. In the center of the home page are many links: Apollo Missions, Astronomy, Black Holes, Cassini Mission, Comets, Extraterrestrial Life, Hubble Telescope, Meteoroids, NASA, Solar System, Space Pictures, Space Shuttle, Space Station, Space Craft, Stars and Constellations, and Telescopes. This is useful for students and teachers. I spent over an hour looking at this site and the cool links.

Title: Cosmology: Astronomy Chapter 26
URL: http://www2.smumn.edu/FacPages/astronomy/p31026pwrpnt/ sld001.htm
Grade Level: High school and above
Best Search Engine: www.askjeeves.com
Key Search Words: what is cosmology?
Review: This page is part of the St. Mary's University's astronomy Web site. The page is part of the presentation from the textbook. However, the text link was down when I tried to view it, so the reference for the text is unknown to me. The site is labeled Cosmology and gives a list of all the pages in the site in an outline forms. The site can be viewed in two ways. You can view it as a graphic presentation or as text slides. If you click on the icon labeled "click here to start," a slide show is presented. To move through the slide show a list of buttons is located on the right hand side of the screen. The buttons look like the buttons on a tape recorder. If you hit the button for play you can move to the next slide. Each slide has a bolded topic. For example, slide 3 is titled "What is cosmology?." The slide has bulleted information about the main topic. Each slide follows the outline presented on the first page. This page is good for high school or college students who want to learn the basics of cosmology. While this page in particular is rather boring to look at, the home page that it comes from has beautiful pictures taken from the Hubble telescope. It also has a list of what looks like an entire semester's worth of astronomy work. This

would be excellent for anyone interested in astronomy. It would also be helpful to high school or college students studying astronomy or college professors or TAs looking for teaching aids or information.

Title: Cosmology: The Study of the Universe
URL: http://map.gsfc.nasa.gov/m_uni.html
Grade Level: High school and above
Best Search Engine: www.msn.com
Key Search Word: cosmology
Review: Cosmology: The Study of the Universe is a straight-forward Web site providing a detailed home page, as well as toolbar options on both the top and left side of the page enabling students to access further information. All general information for this subject can be found from the home page, making it user friendly for students. The first section defines cosmology as "the scientific study of the large scale properties of the universe as a whole." The paragraph goes on to describe more about this science and introduces the prevailing theory, Big Bang. Here it describes observational tests, limitations, and extensions to this theory. Within this top menu are accessible tabs giving the student accessibility to gain a broader search through areas such as the Universe, an Image Gallery, and even a Map Missions Guide. The side toolbar includes topics such as "Big Bang Theory," "Big Bang Tests," and "Beyond Big Bang" for students looking for information on this theory. Overall, I think this site is a great one for any student looking to do research on the topic of cosmology. I think it is a bit advanced for the middle school level but appropriate for a high school–age group.

Title: Cosmic Voyage Museum
URL: http://www.cosmicvoyage.org
Grade Level: Middle school to adult
Best Search Engine: www.metacrawler.com
Key Search Word: cosmology
Review: The Cosmic Voyage Museum home page is well organized. It offers you the opportunity to become a member and receive free news and updates, as well as a free silver-plated key light (pretty cool). I'm not sure what other advantages you would receive if you were a member, so I declined. However, the Web site offers many other opportunities for information even if you do not become a member. If you click on the "Cosmic Clock" you can learn about the evolution of our cosmos through a wondrous journey through time. This section takes you through the first three minutes of time, space, matter, and energy

and explains their theory for it all being born. Also, off of the home page you can click on "Ancient Cosmologies" and learn about the culture of the Egyptians or the lost civilizations of the Maya and see how they viewed the starry heavens. The Cosmic Voyage Museum home page offers a section for the latest news, like new planets that have been discovered and intergalactic cobwebs. At the bottom of the home page there is a software package that you can buy right off of the Web site. It is called "Starry Night Backyard" and it will help you learn all about the night sky.

Title: Ned Wright's Cosmology Tutorial
URL: http://www.astro.ucla.edu/~wright/cosmolog.htm
Grade Level: Grade 8 and up
Best Search Engine: www.excite.com
Key Search Word: cosmology
Review: At my first glance at this Web site, all I can say is "Wow!" Now, Ned Wright may very well be a terrific person, but I must question the life of the man who has put so much time and devotion into this Web site. This site is packed with information. Everything you would ever want to know (and even some things you wouldn't) about cosmology is contained in this Web site. The first thing on Ned's list is a section titled "News of the Universe." At this particular section we can find Ned's latest findings and news clippings on cosmology. Sorting through a lot of stuff I couldn't understand, I found some interesting news on the Hubble, the expansion of the universe, and a new age for the universe that was recently found. The next section of the Web site is entirely devoted to frequently asked questions. This is where you will most likely learn the things you really want to know about cosmology. Ned explores and answers questions on such topics as big bang, red shift, black holes, stars, and dark matter. Next we can move onto the Cosmology Tutorial. In this section, Ned takes you through every aspect of cosmology. It's divided into six separate sections: Observations of Global Properties, Homogeneity, and Isotropy; Many Distances; Scale Factor, Spatial Curvature; Flatness-Oldness; Horizon and Inflation; and Anisotropy and Inhomogeneity. This tutorial is great. If you can stick with it, you can learn a lot about the universe and physics. If your head is still working after the tutorial, Ned also has some interesting things to say about Cosmological Fads and Fallacies, Cosmology and Art, and Cosmology and Religion. This site is definitely worthwhile for most high school students, college students, and adults with an interest in cosmology.

Currents

Title: Ocean Currents—We All Go with the Flow
URL: http://seawifs.gsfc.nasa.gov/OCEAN_PLANET/HTML/
    oceanography_currents_1.html
Grade Level: Middle school and above
Best Search Engine: www.metacrawler.com
Key Search Words: ocean currents
Review: This site is the continuation of the Ocean Planet Exhibit at the
    Smithsonian. The home page offers an ocean topography map and a lot
    of information about ocean currents and other related topics. The page
    offers definitions of the following: Upwelling, Deep Water, Warm Sur-
    face Currents, Western Boundary Currents, and Cold Surface Cur-
    rents. Descriptions of various currents around the world are also given
    (Somalia, California). Upon scrolling down through the home page
    the user will find a link to many Smithsonian lesson plans, including
    one on ocean currents which would be useful in the classroom (or out-
    side of it, for that matter). Underneath this link you will find the link
    to all of the information from the actual exhibit at the Smithsonian
    (Ocean Planet Exhibition Floorplan). I thought that this site was use-
    ful and well organized, but I would have liked to see more links to other
    sites about currents. More interesting about this site, I thought, is the
    connection to the exhibit at the Smithsonian. I think this site would
    be a good resource for children and their teachers or parents.

Title: Ocean Currents
URL: http://ads.x10.com/about/about7_wht23.htm
Grade Level: Grade 6 and up
Best Search Engine: www.metacrawler.com
Key Search Words: ocean + currents
Review: This Web site was laid out in an orderly fashion. It was easy to
    navigate through the site and easy to get back to the home page. If
    you click on Ocean Currents it will give you a listing of the 17 major
    ocean currents of the world, their location, and relative temperature.
    Then click on A Primer on Ocean Currents for an overview of basic
    properties of currents, including velocity, water masses, salinity, and
    much more. Then click on Currents. This will give you a great map
    and a basic overview of the currents of the oceans. Then click on The
    Sea: Ocean Currents. This shows a map pinpointing the location of
    the 17 major currents, with a short description of each. This site also
    has a sidebar with many related topics that you can search or click on.

Title: Currents, Waves & Tides
URL: http://mbgnet.mobot.org/
Grade Level: Middle school and high school
Best Search Engine: www.dogpile.com
Key Search Word: currents
Review: This Web site is maintained by The Evergreen Project and provides the reader with excellent information on currents; specifically on ocean currents. What causes currents and the names to some of the largest currents that exist such as the Gulf Stream are some of the questions that the site answers. The site explains how food and goods from far away lands are brought to us through currents. On the right side of the site there is an animated illustration demonstrating the various types of winds that affect the oceans' currents such as the westerlies and the trade winds. On the left side of the site there is additional information on waves and tides. Another great aspect to this Web site is that it provides the reader with an activity that they can perform on currents, waves, and tides to help them better understand the information provided in the Web site. All in all, this Web site is short but precise with the information provided on currents. The information is easy to understand, especially for students in middle or high school. The information is followed by illustrations and activities, making the information even easier to understand and grasp. Overall, I think the site would be great for students looking for a quick reference on currents, as well as the related topics, with enough information to give a clear and complete picture on the topic.

Title: Ocean Currents "What Goes Around Comes Around"
URL: http://oceanworld.tamu.edu
Grade Level: Grade 4 and up
Best Search Engine: www.google.com
Key Search Words: elementary earth science, currents
Review: The topic of this Web site is currents and ocean circulation. The first section of this topic provides background information for teachers. The Web site discusses five major themes. They are: Systems & Structures, Energy, Change, and Interactions and Measurement. Each topic is discussed in a paragraph. Student outcomes are given for the lesson, as well as a list of resources. Full color graphics are included in the text. The text also contains terms in red that are linked to a glossary. This site is easy to read and could be appropriate for students as young as third grade. The language is appropriately technical. A lot of information is packed into a few paragraphs per topic,

which requires accomplished reading skills and serious concentration. The links to the glossary and the other references within the site are helpful and not so numerous as to be distracting from the flow of the topic information being presented. Teachers would be the main audience for this site, although since it is easy to read the material could be reproduced to be used as text.

Title: Geography
URL: http://geography.about.com/cs/oceancurrents/index.
    htm?iam-metacrawl_1&terms = %2Bwater + %2Bcurrents
Grade Level: All grade levels
Best Search Engine: www.metacrawler.com
Key Search Words: water currents
Review: This Web site came from the about.com Web site. There was much information on currents that is provided with different links to click on to go to different areas. On the left side of the Web site are geography words that you can click on to research geography subject areas. At the bottom of the Web site are links to related sites. On the right of the Web site are news clips that are important for that day that deal with basically anything. The center of the Web site deals with the current section. The links that you can choose in this section are ocean currents, which will provide you information on the 17 major ocean currents of the world; a primer on ocean currents, which tells you an overview of basic properties of currents; currents, which provides a map; and lastly the sea: ocean currents, which pinpoints the location of the 17 major currents. I liked this Web site because it was very easy to find and to move around in.

Title: Pacific Currents
URL: http://www.rhdana.org/pacific.html
Grade Level: Grades 4–6
Best Search Engine: www.metacrawler.com
Key Search Words: currents, elementary
Review: This Web site, available in both English and Spanish, was created by students, teachers, and parents at an elementary school in California, to teach about history, geography, the Pacific Ocean, earthquakes, and volcanoes. When you scroll down the home page, you see a section for sixth-grade Earth Science, with topics like geology, weather, and fossils, and for fifth-grade oceanography class, with subjects like whales, pollution, sharks, and dolphins. You can also click to other marine resource links like aquariums. I clicked on the

Pacific Ocean and got updates on El Niño, the surf conditions, the tide table, and wave heights. This site is extensive, and my only problem would be that if I were a fourth grader, I would have a problem finding my topic because there is so much info given. If you're looking to understand the ocean, visit this site.

Title: Ocean Quest
URL: http://pao.cnmoc.navy.mil/educate/neptune/quest/quest.htm
Grade Level: Grades 4–7
Best Search Engine: www.yahooligans.com
Key Search Words: ocean currents
Review: This site is directly geared toward children between the ages 8 and 13. It was created by the Naval Meteorology and Oceanography Command. Basically, when you first get on the page there are a whole bunch of pictures with captions inside them. You can click on any of the pictures to get information on the ocean floor, sea water, marine life, ports of call, waves and tides, ocean resources, currents, ocean history, and ocean technology. For example, if you click on currents, you are taken to a page that has information on different currents. This link only has one; it is the Gulf Stream. When you click on Gulf Stream, a paragraph comes up explaining what the Gulf Stream is. All of the other headings work in the same manner. If you click back to the main page, there are links on the bottom of the page. From those links you can get to information about the Naval Command, games, links to other pages about the ocean, frequently asked questions, and teacher lesson plans. For students interested in learning about the ocean, I think this is a pretty good page, but if a student had to do a report on just currents, then this would not be the page to choose. There is not enough information on currents to give a student a good grasp of the subject.

Deposition

Title: National Atmospheric Deposition Program
URL: http://nadp.sws.uiuc.edu
Grade Level: Grade 7 to adult
Best Search Engine: www.metacrawler.com
Key Search Word: deposition
Review: This Web site would be an effective teaching tool for the upper middle school and high school student. It was designed and is maintained by the National Atmospheric Deposition Program/National

Trends Network (NADP/NTN), which is a nationwide network of precipitation monitoring sites. The network is a cooperative effort between many groups, including the state Agricultural Experiment Stations, U.S. Geological Survey, U.S. Department of Agriculture, and others. The Web site provides a list of all of the contributors and sponsors. The purpose of this network is to collect data on the chemistry of precipitation for monitoring of geographical and temporal long-term trends. Through the use of this Web site, you can access data products, which include weekly and daily precipitation chemistry data, annual and seasonal deposition totals, mercury deposition data, color isopleth maps of precipitation concentrations, and wet deposition. You can access these products by clicking on the corresponding bolded icon on the home page. In my opinion, a science teacher can use this site to come up with dozens of data retrieval activities for their students.

Title: Sedimentary Depositional Environments
URL: http://www.mines.utah.edu/geo/sedimentology/Sedimentary_ Deposition.html
Grade Level: High school and above
Best Search Engine: www.askjeeves.com
Key Search Words: sedimentary deposition. Note: When "deposition" was used as a key word, most of the links were law sites. Linking deposition and geology was useless, as well.
Review: This site was created by students at the University of Utah to document what they learned on a field trip to the Guadalupe Mountains. The site is personalized by the students. In the introduction two links that state what the students studied are given. They are siliciclastic depositional environments and carbonate depositional settings. Each link gives detailed information and pictures about the topics. Under the introduction is a list of eight links that have a topic and the name of the student who studied that topic. If you point to the link, a picture of the student appears. If you click on the link detailed textual and graphic information is provided about that topic. Below this information is a description of the summer projects that the students studied. The main heading for this information is Carbonate Depositional Environments. Under this heading are seven topics that the students studied. This information is organized like the information provided earlier on the site. Photos of the actual trip are also provided on this page. The site is interesting because it shows the work of actual people doing research. It has a personal feel to it, which makes it appealing. The information that is provided is well

organized and detailed. The pictures and diagrams given in the links give further information on the topics covered in this site. This site is organized in an outline form, which makes it easily navigated. This is a great site for anyone interested in sedimentary deposition.

Title: Paleozoic Deposition and Paleocurrents
URL: http://geology.swau.edu/paleocur/paleoz.html
Grade Level: Grade 3 through college
Best Search Engine: www.yahoo.com
Key Search Words: deposition earth science
Review: This is a no-frills site. It just has pictures of the Earth from different eras and where the land and water levels used to be—which can be interesting and useful with some lessons. How to navigate: Once you are on the home page, you can click on either the picture or the title of the era to enlarge the image.

Title: Deposition
URL: http://www.eagle.ca/~matink/themes/Environ/acidrain.html
Grade Level: All grade levels
Best Search Engine: www.google.com
Key Search Words: deposition and kids
Review: The home page at the top gives you several links: Homepage, Safety Net Newsletter, Teacher Resources, Libraries, Technology, Parents, Just for Kids, Themes, Mini Themes, Holidays, Lesson Plans, and Projects. These links take you to in-depth activities and facts. The home page link takes you to "The Educator's Toolkit." The Teacher Resources button takes you to an extensive list of Special Education Resources. The Library Button gives you the opportunity to use several science search engines. The Themes button takes you to: Weather, Dinosaurs, Solar Systems, Space, Sports, Transportation, and War and Remembrance. The themes were designed for teachers. Back to home page: Under the above list it states: Acid Rain, Table of Contents, Acid Rain, and Lesson Plans. The Acid Rain button gives you many facts about acid rain and the opportunity to explore many fun sites. The Lesson Plan button is built for teachers to create acid rain lessons. This site was well organized and easy to navigate. Each page gave you the opportunity to go back to the home page and has convenient back and forth buttons. However, the site is still under construction. This may get frustrating when you are looking for a specific topic. When it is finished, I think it will be very useful.

Title: Air Pollution and Water Quality: Atmospheric Deposition
    Initiative
URL: http://www.epa.gov/owow/oceans/airdep
Grade Level: Grade 6 and up
Best Search Engine: www.yahoo.com
Key Search Word: deposition
Review: This site comes to us from the Environmental Protection Agency
    (EPA): "This web site is intended to give an overview of atmospheric
    deposition. It discusses the major classes of atmospheric pollutants
    that affect water quality. It also summarizes EPA's efforts to reduce air
    pollution and its impacts on the nation's waters, as well as things you
    can do to reduce air pollution. Finally, we provide links to sources of
    additional information." It is written for the layperson, young adult
    and up and written to help in comprehension of the topic. Important
    scientific terms are in italics to highlight and a glossary is provided
    that gives clear, uncomplicated definitions. It is also possible to use
    the context of the discussion to understand most of the terms because
    of the way it is written. The Main page has eight buttons and a short
    description of each. There is an Introduction, short and sweet, fol-
    lowed by pages (buttons) for Air Deposition, Pollutants Effects, Work
    Plan, Handbook, Glossary, and Information. Each section does a
    good job in giving information and the opportunity to get more in-
    depth by offering other locations for information. The extreme left
    column of the Web site (outside of the Deposition page) offers other
    EPA pages on oceans and pollution. This is a good site for students or
    a layperson looking for information. It is also a resource for profes-
    sionals in need of more detailed information.

Title: Acid Rain and Deposition
URL: http://royal.okanagan.bc.ca/mpidwirn/atmosphereandclimate/
    acidprecip.html
Grade Level: Grade 4 and up
Best Search Engine: www.excite.com
Key Search Words: deposition science
Review: After searching for a good half hour, I finally came across a site
    that had something to do with deposition. This site is very well or-
    ganized. It starts off with an introduction explaining the difference
    between acid deposition and acid precipitation. From my understand-
    ing, deposition is the deposited acidic pollutants from the atmosphere
    on the Earth's surface. Acid precipitation is the pollution that occurs

in rain, sleet, and snow. The site then goes on to explain acid deposition formation. This deposition is normally caused by the pollutants of nitrogen oxides or sulfur dioxide that are released into the atmosphere. The next section of the site is titled "The Effects of Acid Deposition." This section highlights the negative effects of acid deposition to the environment. The site illustrates how fish, vegetation, and humans are all negatively affected by this phenomenon. Toxic metals can be released into the environment through the acidification of soils. Alzheimer's disease has been recently linked to the toxicity of aluminum. The site concludes with a section on solutions to the problem of acid deposition. For lakes that have been found to have high acidic levels, we can now adjust the pH by using a technique called liming. The most obvious solution, however, is to limit the number of pollutants that we are releasing into the air. Factories are already taking steps in the right direction by installing "scrubbers" to limit the amount of sulfur dioxide produced by their smokestacks. In conclusion, I found that this site was very informative about deposition. I learned not only what it is, but also about its environmental effects. A great site for any tree-hugging environmentalist like myself.

Title: Glaciers and Glaciation
URL: http://www.zephryus.demon.co.uk/geography/links/glac.html
Grade Level: Grade 5 and up
Best Search Engine: www.metacrawler.com
Key Search Words: glacial deposition
Review: This page offers eight links that allow the visitor access to "Features of Alpine Glaciation," "Niche and Corrie Glaciers," "Glacial Deposition and Depositional Landforms," "Glacial Erosion and Transportation of Sediment," "Ice Caps," "How Glaciers Form and Flow," "South Pole Exploration," and "Valley Glaciers." "Features" contains drawings of an Alpine glacier and characteristics found in upland glaciated regions and hanging valleys. "Niche and Corrie," gives a simple explanation of the types of glaciers and discusses how each is formed. This link also offers some exciting pictures of famous glaciers, such as the Striding Edge. "Glacial Deposition" briefly discusses ways in which glaciers deposit material elsewhere as well as introduces a few features that may be produced, such as moraine, eskers, and drumlins. The "Glacial Erosion and Transportation" link also gives definitions and characteristics of glaciers defining erratics, till deposits, ribbon lakes, and roche moutonnee. "How Glaciers Form and Flow" link is chock full of pictures as well as an explanation of

how fresh snow becomes glacial ice and the difference between warm and cold glaciers. "Ice Caps" is another wonderful link that offers pictures of drained glacial lake floors and of the ice mass flowing down onto a flat valley floor. The "South Pole Exploration" link offers a paper about the Antarctic written by a doctor stationed at the Amundsen-Scott Station. And, finally, "Valley Glaciers" offers a few pictures of the Mer de Glace glacier in France. Almost all of the links found on this page are written for a school audience. If you can't find it here, you won't find it anywhere!

Droughts

Title: Drought—Discovery.com
URL: http://www.discovery.com/news/features/drought/drought.html
Grade Level: Middle to high school
Best Search Engine: www.yahooligans.com
Key Search Word: drought
Review: I absolutely loved this page. It is done in a professional manner and it is really easy to navigate. The page was created by the Discovery Channel in 2000. The first page you enter has a brief description of droughts and a scary voice comes over the computer speaker and says "drought" like he desperately needs a drink of water. From the main page you can click on the heading "All About Drought." On this page there is more information about droughts and how the National Weather Service can predict when they are going to happen. There is an example of the computer they use that alerts them of drought conditions. In this section there are also pictures of land devastated by drought. The next section to click on is "Famous Dry Spells." When you get there you can click on any of the four headings: The Anasazi, The Lost Colonies of Jamestown and Roanoke, The Dust Bowl, and The East Coast. Each heading shows a short movie on these horrible droughts. When you go back to the main page there is a spot that has links to other pages that discuss droughts, and there is another movie to watch on historical droughts. I think this page would hold a student's attention and give him or her the information they need to know. I would recommend this page to anyone looking for information on droughts.

Title: National Drought Mitigation Center
URL: http://enso.unl.edu/ndmc
Grade Level: Upper middle school to adult

Best Search Engine: www.metacrawler.com
Key Search Word: drought
Review: This site is sponsored by the National Drought Mitigation Cen-
    ter. The center helps people and institutions develop and implement
    measures to reduce societal vulnerability to drought. The NDMC,
    based at the University of Nebraska–Lincoln, stresses preparation
    and risk management rather than crisis management. At the top of
    the home page there are links to the following information: Quick
    Info for Media, U.S. Drought Monitor, Interim National Drought
    Council, and National Drought Preparedness Act of 2002. These
    links provide connections to an abundance of information about
    droughts and everything affected by them all over the world. Under-
    neath, the user will find links and pictures to various topics, such as
    Drought Watch, Drought Science, Climatology, and Methodologies.
    When I clicked on one of the picture links, Methodologies, many re-
    sources were offered to plan for droughts. The information in each
    link is so immense I could not possibly cover it all in this review. but
    take my word you will find it somewhere on this site. I thought this
    site was good but overwhelming. The information is well organized,
    but I suggest the user should have a direct focus before beginning to
    search. Since the information seemed fairly complex I feel that the
    site would better serve older students and their teachers and parents.

Title: Powers of Nature—Droughts
URL: http://www.germantown.k12.il.us/html/droughts1.html
Grade Level: Grade 6 and up
Best Search Engine: www.metacrawler.com
Key Search Word: droughts
Review: This page is part of a site created by a couple of science teachers
    that features a variety of pages dealing with different types of natural
    disasters. At the bottom of this page is a series of text links that take
    the visitor to the main menu, the credit page, or any of the other
    topic pages listed simply by clicking the text. Other topics available
    include earthquakes, floods, slope failure, and many more interesting
    subjects. The page provides a complete background on droughts, the
    three general types of droughts and their definition, and a few text
    links to pages that provide more in-depth exploration of the topic
    (unfortunately, the links were either unable to be found or the infor-
    mation was no longer available when I reviewed the page). The re-
    mainder of the page and text give a brief yet comprehensive look at
    droughts, their severity, and some of the worst cases throughout his-

tory. At the bottom of the text near the bottom of the page, there are two text links that linked to pages that no longer posted the information. Overall, the site is good for students seeking information on droughts; it has solid background information and relevant examples for the visitor to learn about the subject. Unfortunately, most of the links on the page are no longer active or the information at the site the visitor is taken to is no longer available. The site is a good reference to find a quick and viable explanation of droughts, but it lacks a current set of links for further exploration of the topic. The other links within the site will be helpful to anyone seeking information of similar types of natural disasters and their effects.

Title: NOAA's Drought Information Center
URL: www.drought.noaa.gov
Grade Level: Grade 8 through high school
Best Search Engine: www.metacrawler.com
Key Search Word: drought
Review: At the home page there are sections labeled Current Information and Background Information. On the right side there are sections on News (nationwide drought information) and State Drought Information; click on any state for up-to-date drought information. Some examples of the topics in Current Information are U.S. Drought Monitor—drought conditions across the country with a map. There are also "animated" maps to click on in this section, followed by an in-depth National Drought Summary. There is a Seasonal Drought Outlook, updated monthly with a detailed map and seasonal assessment. Other sections are Hazards Assessment and a Drought Calculator, which calculates the amount of rainfall needed to end droughts around the country. In the Background Information section I clicked on All About Droughts. This section answered the question—what is a drought and broke the discussion into several sections, such as Meteorological, Agriculture, Hydrological, and Socioeconomic. There is a section on the impact of droughts on the economy, society, and the environment. The last section describes how meteorologists predict droughts. There are other topics in Background Information such as All About Heat Waves, Climate 2002, Climate Monitoring (weekly), Climate at a Glance, State Climate Offices, and Regional Climate Centers. At the bottom of this page is contact information with more government sites on droughts. In summary, this site is well written and easy to navigate, and it provides detailed information. There are a lot of maps and graphs.

Title: Drought: The Creeping Disaster (Earth Observatory)
URL: http://earthobservatory.nasa.gov/Library/DroughtFacts
Grade Level: Grades 7 and up; may be challenging for pre–high school
Best Search Engine: http://www.google.com
Key Search Word: drought
Review: On the right side of the page is a quote capturing the main idea
   of the section and an outline of the sections of the site: Introduction,
   Drought Planning and Drought Indices, Improved Monitoring, and
   Improved Forecasting, which link to the specific section. In addition,
   there are links listed to Related Links within the larger Earth Observatory parent site: Drought and Vegetation Monitoring and Measuring Vegetation (NDVI and EVI). At the top of the page above the
   title of the topic, you can choose to turn the glossary on or off. When
   on, words are highlighted in red from which you can link to a glossary
   displaying that specific term only, with options either to select a letter of the alphabet or to view the complete glossary. The glossary is
   for the whole Earth Observatory parent site and is comprehensive.
   The drought site contains effective color graphics. The topic is presented one section at a time, with the option at the end of each section to link to either the next section or to the previous section.
Site Review: This is a very informative site. It not only defines drought by
   three criteria, but it also gives a recent historical perspective of the
   impact of drought, explains how many types of data are collected and
   assimilated, and discusses advances in drought monitoring and forecasting. All key points are effectively illustrated by the color graphics.
   The language is appropriately technical, perhaps a bit difficult for
   some pre–high school students, but the glossary is helpful and it is an
   interesting presentation, not at all dry (sorry, I couldn't resist). The
   site is well organized and easy to navigate. I especially liked the bal-
   anced coverage of the subject, which in my opinion made it absorb-
   ing.

Title: Science Benchmark Clarification, Instruction and Assessment
URL: http://www.miclimbscience.org/docpieces/st5d_frame.html
Grade Level: Grade 1–6 teachers
Best Search Engine: www.metacrawler.com
Key Search Words: droughts, elementary
Review: This Web site is part of the national science education standards
   and deals with using science knowledge in real-world contexts. This
   content standard says that all students will be able to describe the
   Earth's surfaces, explain how the surfaces change over time, and ana-

lyze the effects of technology on the surface and Earth's resources. It gives examples of lesson plans you can do with the various grade levels using the key concepts in real-world contexts. It also gives additional related links to access for more ideas. Included in the plans are the rubrics, materials needed, photos, and so forth. It's a wonderful site for teachers who are stuck for easy lesson plan ideas . . . fun for kids to look at.

Title: All About Droughts. . . .
URL: http://www.nws.noaa.gov/om/drought.htm
Grade Level: Middle school and up
Best Search Engine: www.dogpile.com
Key Search Word: droughts
Review: This Web site supplies the reader with helpful and a significant amount of information on droughts. What is a drought is the first question answered within the Web site. The question is answered in great detail and includes the four ways a drought can occur: meteorological, agricultural, hydrological, and socioeconomic. These four types of droughts are defined in a way that a middle school or high school student will understand. What are the impacts of a drought is another question that is answered throughout the site. The site lists the three types of main impacts: economic, social, and environmental. Each of these impacts can be clicked on so that the reader may obtain further information. This further information is found at the National Drought Mitigation Center (NDMC) Web site. One of the great aspects of this Web site is that it provides additional sites that readers can visit to further educate themselves on the topic of droughts. The National Climate Data Center and the Federal Emergency Management Agency are just some of the additional Web sites provided.

Title: Drought
URL: http://www.discovery.com/news/features/drought/drought.html
Grade Level: High school
Best Search Engine: www.yahoo.com
Key Search Word: droughts
Review: This Web page on droughts was very informative. Included is a link that you can click on to see a video of an area where there is a serious drought. Also included in the Web site is a definition of what a drought is and how serious one is to land. At the bottom of the page is a picture of a drought. To the right of the page are related links that

can help you find more information on droughts. I liked this page since a reader can see a picture of what is being described. I also felt it was very easy to view.

Title: NOAA'S Drought Information Center
URL: http://www.noaa.gov/
Grade Level: Grade 4 to adult
Best Search Engine: www.google.com
Key Search Words: earth science lessons, elementary, droughts
Review: This is the site of the National Oceanic and Atmospheric Administration. It contains valuable information on a variety of weather related topics. Listed on the main page are a variety of articles that you can access by clicking on the links provided. Further down the page, you can click on such topics as Weather, Ocean, Satellites, Fisheries, Climate, Research, Coasts, and Charting & Navigation. Listed underneath is a section called Check this out . . . , which is a list of interesting articles. The next listing is titled Cool NOAA Web Sites. Here you can check out a variety of sites offered by NOAA on a variety of topics. The last listing is the Media Advisories section. On the right is a column with such listings as the NOAA seal and today's date followed by the question of the month. The next listing is for NOAA's magazine. The next section in the column is titled About NOAA. Here you can find links that provide a variety of information about the agency. Listed in the next section is a variety of topics such as Image of the Day, High-Resolution Satellite Images for the Media, Storm Watch, Weather Forecasts, U.S. Hazards Assessment, Drought Information Center, Excessive Heat Index, Heat Wave Index, Severe Weather Awareness, Hurricane Awareness, Fire Weather Information Center, Fire Satellite Images, Current Satellite Images, Archived Images of Significant Events, and News Story Archive. I would choose the link entitled Drought Information Center for information on droughts for this review. This is NOAA's drought information center. It is a roundup of the various NOAA Web sites and information on drought and climate conditions. Some external links are included for your convenience. These are the links listed on the Drought Information Center page: Current Information U.S. Drought Monitor—(assessment of recent conditions and drought status) (A joint effort between federal and academic partners.). Seasonal Drought Outlook from NOAA's Climate Prediction Center—(updated monthly). Hazards Assessment (extreme weather conditions—with graphics). NOAA's Drought Assessment—includes the latest graphics Drought

Calculator—NOAA calculates amount of rainfall needed to end droughts around the country. Click here to go directly to NOAA's Drought Termination and Amelioration. U.S. Soil Moisture Monitoring. NOAA's Drought Monitoring. Palmer Drought Severity Index— (graphic updated weekly). All About the Palmer Index. Palmer Drought Information by Region Crop Mois-ture Index—(graphic updated weekly). Background Information. All About Droughts. All About Heat Waves. Billion Dollar Weather Disasters—1980–1999. Climate of 2002—U.S. and Global Climate Perspectives. Climate of 2001—Annual Review. Climate Monitoring—Weekly Products. Climate Monitoring—Reports and Products. Climate At A Glance— Nationwide Temperature & Precipitation. Extreme Climate and Weather Events. Fire Potential. National Geophysical Data Center— Drought Variability. National Climatic Data Center—Climate Division. Drought Data—Graphing Options. NOAA's Climate Theme Page. NOAA's Climate Diagnostics Center. Drought Monitoring— Current and Anticipated Precipitation Anomalies over the U.S. NOAA's National Centers for Environmental Prediction. NOAA's Office of Hydrology. State Climate Offices. Regional Climate Centers. North American Drought: A Paleo Perspective. U.S. Geological Survey Water Watch—Maps and Graphs of Current Water Resources Conditions. National Drought Mitigation Center—(USA clearing house for drought information). Vegetation and Temperature Condition Index. This site also integrates other disciplines into the study of droughts. It also provides news on droughts and has a link for several states to check drought conditions. The last listing on this page is other links which contains a variety of subjects such as Archived News Releases NOAA Home Page, NOAA Public Affairs, NOAA Media Contacts. On the home page there is a listing of related organizations and topics. Also listed on the home page and worth noting are areas for opportunities, such as employment and education. I believe that this is the most thorough and informative site there is on any weather-related topic. It contains something for every age group.

Earthquakes

Title: Earthquakes
URL: http://www.fema.gov/kids/quake.htm
Grade Level: Grades 4–6
Best Search Engine: www.metacrawler.com
Key Search Words: earthquakes and kids

Review: This colorful and informative Web site is a great resource for students and teachers. The site has rumbling sound effects, colorful images, and easy navigation. Earthquakes are defined in a clear and easy to understand manner. Students are introduced to the Richter scale and can read what the ratings on the scale mean. The site is full of great links such as a fact and fiction page. Students can click on this page to find out if myths about earthquakes are correct. Located vertically on the left side of the home page is a colorful index. The topics are: Shake with the Quake Story (cute), Rumble Tumble Story (fun), The North Ridge Earthquake, Fact or Fiction (fun), Home Hazards Hunt, Historic Earthquakes, Tasty Quake (silly science activities), Map of Earthquake Risk (informative), Earthquake Disaster Math (little test), Disaster Intensity Scales, Water Wind and Earthquake Game (interactive), Earthquake Legends (global stories), and Jess and Sam's Earthquake (three fun experiments.) The site also allows kids to link to pages that discuss historic earthquakes and to view a map of earthquake risk states. The site also has a game that students can play. There are also tips for how to stay safe during an earthquake. This is a great site that is appropriate for upper elementary school and can be a great teaching tool for educators. This site was created by FEMA (Federal Emergency Management Agency). It was my favorite site that I researched. As a teacher, I found the site was interdisciplinary and informative. As a kid, I found the site fun!

Title: Understanding Earthquakes
URL: http://www.crustal.ucsb.edu/ics/understanding/
Grade Level: Middle school to high school
Best Search Engine: www.yahooligans.com
Key Search Word: earthquakes
Review: There are six links on the main page of this Web site. Earthquake Quiz is an interactive quiz to test your knowledge of earthquakes. Rotating Globe is a JavaScript-enabled page that enables you to locate earthquakes around the world. Famous Earthquake accounts tells of the earthquake experiences of Mark Twain, Jack London, Charles Darwin, and John Muir. How Earthquakes Occur is another Java animation page. When you click on the picture, you are shown how earthquakes begin. There is also a lengthy description of how earthquakes happen. The History of Seismology offers three pages of information starting in 1910. The last link offers 14 additional earthquake Web sites, including two from the U.S. Geological Survey. One of the sites offers an up-to-the-minute map featuring California earthquakes. This is also a

JavaScript site. The link to the Earthquake Engineering Research center offers many historical photographs of earthquakes. This is a good site for teachers to introduce earthquakes to students. The animated pictures are helpful in showing how an earthquake occurs.

Title: The BGS Earthquake Page
URL: www.earthquakes.bgs.ac.uk
Grade Level: Grades 9–12
Best Search Engine: www.yahoo.com
Key Search Word: seismology
Review: This is an information-filled Web site about the earthquakes in Britain. There is a booklet that can be downloaded and printed. There are lots of links that discuss everything about earthquakes in Britain. There is a list of recent earthquakes that have occurred in the past 30 days. There is a section on information and facts about earthquakes. They answer common questions such as what an earthquake is and where they occur. They also include a glossary of terms, references, educational links, and much more. This site also discusses the felt effects of earthquakes on people, buildings, and nature. It teaches how earthquakes are monitored and weighed. There is also a list of other links associated with earthquakes around the world. This is a great site for those studying earthquakes from around the world, not just in the United States.

Title: Plate Tectonics, the Cause of Earthquakes
URL: http://www.seismo.unr.edu/ftp/pub/louie/class/
    100/platetectonics.html
Grade Level: Grade 4 through college
Search Engine: www.yahoo.com
Key Search Words: earthquakes, earth science
Review: A site with succinct information concerning "how, why, and where" with great pictures. How to navigate: once you are on the homepage, just scroll down. It isn't until you get to the very bottom of this site that the reader learns that this is part of a lecture series on Earthquakes. It explains a lot. There is a fair amount of semitechnical jargon and a lot of photos that illustrate the text very effectively. This is a large Web page with everything on plate tectonics available as you scroll down the page. The photos are spectacular. Different areas of the globe are highlighted by different satellite photos all to illustrate the particular definition/segment under discussion. Beautiful. The first picture is actually a map of the Earth that plots where the

seismic activity has been charted. It very clearly illustrates where the plates are located just by the dots denoting where earthquakes have been recorded. Once the reader has finished digesting the information on the page there are two links at the bottom of the page; one for the next lecture on Seismic Waves and another for "more information about Earthquakes" that leads to a page with a variety of new information and the balance of the lecture series topics.

Title: What Are Earthquakes?
URL: http://scign.jpl.nasa.gov/learn/eq.htk
Grade Level: Middle school and above
Best Search Engine: www.askjeeves.com
Key Search Words: What is an earthquake?
Review: This site is part of a home page that was developed by the Southern California Integrated GPS Network Education Module. The page gives a brief definition of what an earthquake is. It gives an example of a pencil that is being pushed at both ends. When enough force is exerted, the pencil will snap. At the bottom of the page the reader can push buttons that put force on a pencil icon. When the reader pushes on the button marked "push the ends of the pencil" the pencil icons bends. When the "push the ends of the pencil harder" the pencil icons breaks. This is a simple visual representation of what occurs in an earthquake. Along the left hand side of this Web page are 11 other links. These links are labeled Home, Overview, Plate Tectonics, Earthquakes, GPS, Space Technology at Work, Activities, People in SCISN, Glossary, Comments and Suggestions, Other SCES Modules, and SCIGN Homepage. These links provide various types of information. The topical links give information similar to the information on earthquakes. The other links give information on how educators and students can use the site and the information on the creators of the site. This site gives a simple description of earthquakes. It does not provide detailed information or detailed graphic information. While the site states that it is for high school students and college undergraduates, I think that middle school children could read and interpret the information provided. The graphic information on the site may be boring for high school students.

Title: Understanding Earthquakes
URL: http://www.crustal.ucsb.edu/ics/understanding
Grade Level: Grade 3 and up

Best Search Engine: www.excite.com
Key Search Words: earthquake education
Review: This was the first site I found and I was quite pleased with the results as soon as the page loaded. This is an educational site for either students or teachers that gives a lot of really great information on earthquakes. The main page has different links to choose from. These are Earthquake Quiz, Rotating Globe, Famous Earthquake Accounts, How Earthquakes Occur, History of Seismology to 1910, and Other Educational and Earthquake Web sites. The earthquake quiz is fun. It's just four questions (and answers) to test your knowledge about earthquakes. The rotating globe section is an animated globe that shows the locations of earthquakes. Famous earthquake accounts lists some written accounts of earthquakes documented by Mark Twain, Jack London, Charles Darwin, and John Muir. How earthquakes occur is exactly just that. An animation shows the gradual buildup of stress that leads to an earthquake. The history of seismology is just a three-page account of the history of this branch of science. A little dull, but a necessary read for the subject of earthquakes. And finally the last section is links to a multitude of other earthquake sites. Overall, I felt that this Web site serves as a great educational introduction for students when teaching earthquakes. Teachers should not hesitate to use this site in class. The animations alone are worth the visit.

Title: The U.S. Geological Survey
URL: http://www.earthquakes.usgs.gov/4kids
Grade Level: Grade 6 and up
Best Search Engine: www.msn.com
Key Search Words: earthquakes, science, children
Review: The U.S. Geological Survey Web site is a great place for younger children to learn many facts about earthquakes. This site has many areas geared toward elementary school children. Brightly colored, the home page draws your attention to the many separate boxes which you can click on to access the different pages in this site. You can not only learn a variety of information as to the causes of earthquakes and what they are, but also this site has an area that teaches children the hazards of earthquakes and how to prepare themselves should an earthquake occur. The many areas in this site include "Latest Quakes," where children can see where recent earthquakes have taken place; Earthquake Image Glossary, where the children can see the aftermath of the earthquake; and Today in Earthquake History. Online activities and Science Fair Ideas and Projects are also avail-

able, which add to the fun approach to learning. The area Cool Earthquake Facts tells many random facts about quakes and includes pictures and diagrams to help the children visualize. This site also contains links to regional Web sites and various seismic networks for further research.

Elements

Title: The Weather Doctor—Weather Phenomenon and Elements
URL: http://www.islandnet.com/~see/weather/doctor.htm
Grade Level: Grade 6 and up
Best Search Engine: www.metacrawler.com
Key Search Words: elements of weather
Review: This site is jam-packed with information for anyone seeking information of any type on weather or weather-related topics. The main page welcomes the visitor to the site and provides a scrolling screen on the right side of the page that covers current news and information posted on the site, as well as a variety of interesting general topics. Along the left side of the page is a text link index of topics available to the visitor. It is in this list that we find the topic Weather Phenomenon and Elements. This links the user to a home page for the topic while remaining in the same familiar navigation, although the new subject pages now have a new list of links on the left side that are subtopics of the subject and the right side hosts the content that the user clicks on from the left. The entire site maintains a top navigation that remains constant and links the user back to the site's main page, allowing the user to never get completely lost or far from home—especially with a link to the site map. The page we are examining that features information on weather phenomenon and elements has a comprehensive listing of topics that includes When Hard Rain Falls, Making Clouds and Rain, Thunder, Wind, and many, many more. Each text link leads to a new page that is loaded with definitions, explanations, charts, maps, animations, and much more to help explain the topic and provide as much useful and relevant information as possible. The main weather phenomenon and elements page also contains a link to a Weather Almanac Index, which provides further information and discussion on the elements. Overall, this site is extremely complete and offers an awesome amount of detailed information on this topic. The pages are loaded with extensive descriptions and visual examples. I think it is a terrific site that offers a number of subtopics related to the elements that would provide any

student with plenty of information to better understand the topic of elements.

Title: Periodic Table of the Elements
URL: http://pearl1.lanl.gov/periodic/default.htm
Grade Level: All grade levels
Best Search Engine: www.yahoo.com
Key Search Word: elements
Review: This is a colorful Web site. At the top of the page there is a sample of the periodic table of elements that are in all different colors. Under the periodic table is two sections of other elements. These elements are in the Lanthanide Series and the Actinide Series. Following this section is an area where you can download the periodic table. The last section is different links that you can click on to go to other Web sites that deal with the elements. These include: What is the Periodic Table?; How to use the Periodic Table; Mendeleev's original Periodic Table; Chemistry in a Nutshell; and Naming New Elements. I liked this Web site very much because it was informative. It was also interesting and easy to look at.

Title: kapili.com
URL: http://www.kapili.com/e/elem_hydrogen.html
Grade Level: Grade 4 to adult
Best Search Engine: www.google.com
Key Search Words: earth science, elements
Review: To find this site from the home page, click on topic list, then choose either elements or elements and the periodic table. I chose Elements, which is a brief look at the elements, great for introducing the elements to students who are not yet ready for an in-depth study. Choose one of the headings—Elements 1–10, Elements 11–18, Elements 19–36, Elements 37–54, Elements 55–86, Elements 87–100—and click. Each element will be listed in a column on the left side of the page. Pick the element you would like to study and click. For each element, the Name, Symbol, Atomic Number, and the Atomic Weight is given as well as any important additional information. This is a great site for younger students as well as those who may need a quick refresher course.

Title: Chemical Elements.com, An Online, Interactive Periodic Table of the Elements
URL: http://www.chemicalelements.com

Grade Level: Middle school and high school
Best Search Engine: www.dogpile.com
Key Search Word: elements
Review: This Web site is an interactive periodic table of the elements. This Web site is a great way for one to learn all they need to know about the elements and have fun while doing so. The Web site allows the student to observe the periodic table in numerous ways such as illustrating the element's names, atomic numbers, atomic masses, electron configuration, number of neutrons, melting point, boiling point, date of discovery, and crystal structure. In addition the student has the capability to categorize the periodic table into specific element groups such as Alkali Metals, Alkaline Earth Metals, Transition Metals, Other Metals, Metalloids, Non-Metals, Halogens, Noble Gases, and Rare Earth Elements. At any point within the Web site, the student may click on any element's symbol to receive more information on that particular element. Everything one would want to know about an element is provided within this information such as everything mentioned above, its use and a diagram of its atomic structure showing its number of energy levels, and isotopes. All in all, this site is an informative yet fun site for one to learn everything they want to know about the elements. I think this site is a great "quick reference" for the sciences. However, the K–12 rating was greatly overestimated. Children in grades K–5 will have little use for this information but, I think it could serve as an invaluable resource for older students. The information is quite direct and the wording is simple and clear. Overall, the site is well organized and easy to navigate. I highly recommend www.chemicalelements.com to anyone taking a high school or college chemistry course.

Title: Meteorology Online
URL: http://library.thinkquest.org/C0112425/child_intro_1.htm
Grade Level: Grade 4 (proficient readers) and up
Best Search Engine: www.google.com
Key Search Words: weather elements
Review: Navigating the Site: Across the top of the site from left to right are the following choices: Home, Children, Student, Library, Laboratory, and Staff. The following site tree shows what is included in each of these areas. Site Tree Home—Information—Café—Guestbook Children—Chapters > Introduction > The Sun > Air Pressure > Air Moisture > Air Masses, Fronts > Winds > Severe Weather > Extreme Weather Conditions >Weather Forecast >Weather Maps

> Weather Calendar > Weather God—Resources—Courses > The Greenhouse Effect > Acid Rain > El Niño—Games Student— Chapters > The Sun > Winds > Storms & Fronts > Rain > Thunderstorms & Tornadoes > Sky Watching > Prediction Weather > Future of Earth—Resources—Games Library—Children—Student— Teacher's Corner—Media Clips—Gallery Laboratory—Children > Rain Detector > Cloud > Rain Gauge > Barometer > Rainbow > Thermometer > Air Pressure—Student Staff—Staff Office—Staff Lounge Bibliography Glossary. I selected "Children," and the subcategories shown in the site tree under Children are displayed throughout the tour on the right side of the screen. At the start of each subcategory, an outline of that topic is displayed at the top of the text box on the left side of the screen; for example, Introduction, followed by the section headings What is Meteorology?, What is Weather?, and 5 Major Weather Elements. Just above the text box, there are five buttons: Information, Glossary, Home, Site Tree, and Bibliography. The glossary opens in a separate small window. It contains a comprehensive selection of meteorological terms in simple language for children; for example: "Hurricane—a swirling storm in which wind speeds are above 120km/h." You can either move through the subcategories in sequence as a site tour by pressing the Forward button at the bottom of each subcategory, or link directly to a subcategory from the subcategory menu, which remains displayed at the right of the screen. You can enlarge the color images by clicking on a magnifying glass icon beside the graphical image. Inserted throughout the text are short sections titled, "Did You Know," which contain anecdotal data; for example, "Did You Know? Without the weather to spread the Sun's heat around the world the Tropics would get hotter and hotter and the Poles colder and colder. Nothing would be able to live on the earth." Also occurring throughout the text are icons that link to labs for children, such as "Making your own simple thermometer" as follows: "Making your own simple thermometer. You will need a) a strong plastic bottle with a screw cap b) a thin plastic straw c) some modeling clay d) a cellotape; and e) some water coloured with poster paint. Step 1. Ask an adult to drill a hole in the bottle's cap. Assemble the thermometer, making sure that the water comes part way up the straw when you fix on the cap. Leave the water to settle for an hour then mark the water level on the scale. 2. Stand the thermometer in a bowl of ice cold water and a bowl of very hot water. See how the water level changes." The labs open in the text box, replacing the topic, but after the lab instructions there is no nav-

igational aid to return you to where you left off in the text. You have to use the browser's Back button. This was the only place in the site where the navigation was poorly thought out. Site Review: The Meteorology Online site covers the five major weather elements thoroughly, woven throughout the presentation of the science of meteorology, which even includes what various cultures once believed caused weather ("Weather Gods"). The text is in a large point size, which is easy for younger students to read. The outstanding color graphics make up about one-fourth of the content, again making the reading easier and more enjoyable for younger students. Nevertheless, the text is intelligently written so that an adult who wanted a comprehensive introduction to the subject of meteorology would not be disappointed. I especially liked the site because it is designed for younger students, with straightforward language and plentiful color graphics, yet the presentation is sophisticated and graphically well designed.

Title: Periodic Table of the Elements
URL: http://www.klf-split.hr/periodni/en
Grade Level: Grades 9–12
Best Search Engine: www.google.com
Key Search Words: scientific elements
Review: The home page of this site has a large Periodic Table of the Elements. The table is color-coded to identify solids/liquids/gases. Underneath the table there is a description of how to find the relative atomic mass for each element. There is a number in brackets to indicate the mass number of isotopes for the elements. The next section is Content, which contains a glossary of chemical terms with brief definitions of the elements. Just click on each letter for further information. There are Periodic Table of Elements listed in five languages—English, French, Croatian, German, and Italian. There is more information on the periodic table listed alphabetically by names and symbols. (Just an FYI—this project is financed by the Ministry of Science and Technology in Croatia.) There is a printable periodic table in color and black-and-white. The next section includes scientific calculators for chemists and a molar mass calculator. There is a section labeled Tools and Toys, programs for chemists and nonchemists. There are additional links for chemists and chemistry journals. At the end of the page are statistics and awards for the periodic table and the project. The goals of the project are outlined, and there is information on software for chemistry. This site was well written

and very technical. It was easy to navigate, and there are some really nice charts on the elements.

Title: Periodic Table.Com
URL: http://www.periodictable.com
Grade Level: Grade 4 and up; students, teachers, chemists, biologists
Best Search Engine: www.metacrawler.com
Key Search Words: elementary, elements
Review: The Web site is divided into separate sections for students, teachers, and others. I clicked on the students section and it gave me the option of choosing the periodic table, riddles, other links, or history. I chose riddles, assuming most kids would choose this. I found challenges, riddles, anagrams, jokes, cartoons and fun, but I also learned about the elements in the process. It's a smart idea. Kids probably don't even realize they're working while they're on the site. I do think it's best for older kids, though, maybe fourth grade and up.

Energy

Title: Tropical Rainfall Measuring Mission
URL: http://trmm.gsfc.nasa.gov
Grade Level: Grade 6 through college
Best Search Engine: www.yahoo.com
Key Search Words: energy earth science
Review: Interesting site describing the Tropical Rainfall Measuring Mission, which is a joint mission between NASA and the National Space Development Agency of Japan to monitor and study tropical rainfall and the associated release of energy that helps to power the global atmospheric circulation shaping both weather and climate around the globe. Site offers "breaking weather news," "rainfall measurements," and "see latest 3 hourly global rainfall." All come complete with really neat pictures. How to navigate: once you are on the home page, just click on your desired icon. The site is also searchable.

Title: U.S. Department of Energy
URL: http://www.energy.gov/kidz/kidzone.html
Grade Level: Middle school and above
Best Search Engine: www.msn.com
Key Search Words: energy, science, children
Review: The U.S. Department of Energy has a Web site designed just for kids! The initial page is colorful and pulls up a link to a "featured site"

which changes every month to address a different aspect of energy. There is sidebar at the top left where you can chose options to obtain additional information about the Department of Energy itself, as well as information regarding various forms of energy. At the bottom of this main page are "quick links" where children can access different sites to enter contests, locate events, play games, take quizzes, or find science projects. There is also an area at the bottom right called "Ask Energy Ant" where children can go to type in questions. Children can go to the "About US" link to get a basic description of what the Department of Energy does, and how and why it was created. There is also a selection called the "Energy Dictionary," where students can access such topics as atoms, accelerators, and biopower. Children can also access an interactive Periodic Table of the Elements on this site, which would probably make learning more interesting. Here, they can also link to the Hawaiian Volcanic Observatory. Overall, this seems like a great site for kids to get a complete idea of what energy is, what the various types of energy are, and why energy is so important to human life.

Title: Energy!
URL: www.ippex.pppl.gov
Grade Level: Middle school and above
Best Search Engine: www.askjeeves.com
Key Search Words: energy in physics in askjeeveskids
Review: This page is a subpage of The Internet Plasma Physics Education Experience home page. The mission of this site is to provide hands-on physics information to students and educators in an effort to meet educational standards in the United States. The top of the page defines energy. Then a one-page introduction to energy is provided. Within this introduction are various sites that provide further information that will help the reader understand energy. The links are included within the black text, in red writing. These links take you to various pages that provide definitions of various terms and graphics that give additional information. One of the links has an energy experiment. Along the left part of the page are three main headings with several subheadings underneath. The main headings are Information, Education, and Support. These provide information about the organization that created this site, other topics in physics like fusion or interactive physics, and information on how to navigate the site. This site provides a brief description of what energy is in scientific terms. It also provides links to give the reader a better under-

standing of the physics involved in energy. The site is brief in its content and could have provided more graphics in the initial pages. This is a good site for students trying to get background information on how energy works. The lab provided could be used by educators as well as students.

Title: Destination: Earth the Gateway to NASA's Earth Science
    Enterprise
URL: http://www.earth.nasa.gov
Grade Level: Grade 12 and up
Best Search Engine: www.yahoo.com
Key Search Words: energy and earth science
Review: Navigation: Select Science button at top of page. That takes you
    to http://www.earth.nasa.gov/science/index.html, Science of the
    Earth System (page title). Select Global Water and Energy topic on
    top of the page. This is obviously not an easy page to get to; although
    I deliberately took the long way through the original site because
    there are an awful lot of interesting topics to explore through the
    main page. It is easy to get side tracked getting to the Global Water
    and Energy topic. This subsite is primarily concerned with how the
    climate is created through the cycling of water, water vapor, and heat
    and how the changes in ocean and atmospheric conditions affect
    weather. This page is not for younger students; it is primarily a re-
    source for higher level research as it enumerates a multitude of re-
    search projects, both ongoing and completed. Younger students
    would probably not have the patience to wade through the various
    support sites to find information that would be useful. This would
    probably also be true of the layperson looking for quick information.
    For more in-depth info this would be a good place to start. There is an
    awful lot here.

Title: Energy Information Administration
URL: http://www.eia.doe.gov
Grade Level: Grade 4 and up
Best Search Engine: www.metacrawler.com
Key Search Word: energy
Review: The Energy Information Administration offers tons of informa-
    tion about energy. From the home page the visitor can access infor-
    mation about press releases, historical data, energy links, and energy
    events. The page also offers the user access to information about en-
    ergy by searching "By Geography," "By Fuels," "By Sector," or "By

Price." I decided to look at "By Geography." Here a map of the United States appeared and I was then able to select from the 50 states by the two-letter state code. I looked at Texas and was then able to choose from petroleum, natural gas, electricity, or total energy. Total energy listed prices and consumption, it also offered reports on appliances. In playing around I had located a link called the "EIA Kid's Page." Here Host Energy Ant can help children research questions like "What is Energy?" Energy Ant also lists fun facts about energy, online resources, a classroom connection, and an energy quiz. I clicked on "What is Energy?" and found myself at a page that offered information about renewable and nonrenewable resources. A simple explanation about the different forms of energy were listed such as thermal, radiant, mechanical, electrical, chemical, and nuclear. The definition also discussed the two types of energy, potential and kinetic, in simple terms. The visitor could look at the types of energy we use on a daily basis in our homes, vehicles, businesses, and schools. Because I know so little about the way energy is measured, I found the conversion chart link to be useful. The link helps students convert measurements into a usable format. For example, the student can convert physical units of energy, such as barrels, tons, cubic feet, into BTUs, thus producing a practical way in which to compare different fuels. "Fun Facts" amused me for some time. There I found that about 3,000 products other than gasoline, diesel fuel, and heating oil are made from crude oil. I was surprised to find that ink, crayons, bubblegum, deodorant, eyeglasses, and heart valves are made from crude oil. As an adult I found the children's link informative and fun. As a teacher I would definitely use this site while teaching an energy unit.

Title: Solar Energy International
URL: http://www.solarenergy.org
Grade Level: Grade 3 and up
Key Search Word: energy
Best Search Engine: www.metacrawler.com
Review: Solar Energy International (SEI) is a nonprofit organization whose mission is to provide education and technical assistance so that others will be empowered to use renewable energy technologies. There philosophy is that renewable energy resources can improve the quality of life and promote sustainable development throughout the world. SEI works cooperatively with grass-roots and development organizations to meet sustainable development goals with renew-able energies. This Web site provides access to technology transfer programs, renewable energy

training programs, and decision-making workshops. If you click on the heading "students," you will find the latest news and information being offered from SEI for students of all ages. This site is a terrific resource for teachers who are interested in energy workshops and programs. Teachers can transfer their newly learned information to their students.

Title: Energy Science
URL: http://www.energyscience.co.uk
Grade Level: Ph.D. and above
Best Search Engine: www.excite.com
Key Search Words: energy science
Review: This site is originally divided from the main page into three subsets. You can choose Overview, which is for the general reader, a fascinating account of the mysteries of physical science. You can choose Physics, which is strictly for the physicist. This provides a comprehensive reinterpretation of what has been misconceived in physics. And the third and final section is Technology—for the technologist there is a challenge of gaining access to nonpolluting energy resource of the space medium. Of course I wasn't daring enough to enter the physics or the technology realm at first, but after going through the overview, I ventured in for a visit. I'm not a stupid man, but after visiting this site, I feel like one. The overview section provides an in depth view of the history of energy in physical science. I had no idea what half of it was talking about. After reading pages about aether, I finally found some familiarity with a discussion on nuclear fission: hot and cold. After my mind recovered from the numbness, I ventured to the Physics section. This was actually a little easier to follow. There is a nice welcome message from the author of the site, explaining his accreditations as a physicist, the books he's published, his lectures, and his purpose for the site. The text from parts of his book and his lectures are all laid out there as well. Finally, I took a quick look at the Technology section. This part of the site explained how he wants to find an alternative means to energy through a type of cold fusion. This section also contains essays, lectures, and notes on his theory. Overall, I felt this site was inappropriate for 99.9 percent of the population. Unless you have a Ph.D. in physics, or a longing for a headache, stay away from this site.

Erosion

Title: Dirtmeister's Science Reporters: Investigate and Report on Erosion
URL: http://teacher.scholastic.com/dirtrep/erosion/index.htm

Grade Level: Middle school
Best Search Engine: www.google.com
Key Search Words: erosion and kids
Review: On the home page of this Scholastic Web site you will find the
    Dirtmeister. To the left of his picture there is a list of the current in-
    vestigations; at the top of the list the user will find a link to reports on
    erosion. Animal adaptations, Friction, and Simple Machines are the
    other investigation topics listed. When I clicked on the erosion link I
    was taken to a page titled, "How do the forces of erosion change the
    world in which we live?" Here, the Dirtmeister divides the investiga-
    tion on erosion into four parts or links. They are 1) Investigate the
    Facts, 2) Observe and Record, 3) Report your findings, and 4) Read
    sample reports. The "investigate the facts" link will take the user to a
    fairly detailed information page (five paragraphs) about the topic,
    erosion. This page gives basic facts and pertinent information about
    erosion. At the end you will find a link to the next step, observe and
    record. Users are directed to a page where they are asked to observe
    the effects of erosion in their own neighborhoods. The Dirtmeister
    offers a printout record sheet. This sheet lists guiding questions to
    help the reporter investigate. Next, the user is asked to report his or
    her findings using the example report provided. Link 4 will take you
    to sample reports. This site is great for students in the middle school
    grades because the language is upbeat and fun and the information is
    relevant for them. On the home page one will also find a teachers
    guide as well as a link to the Dirtmeister's hands on Science Labs. I
    think this could be a great resource for teachers and students on a var-
    ied number of topics.

Title: Geoindicators—Soil and Sediment Erosion
URL: http://www.gcrio.org/geo/soil.html
Grade Level: Grade 8 through high school
Best Search Engine: www.metacrawler.com
Key Search Word: erosion
Review: This site is produced by the U.S. Global Change Research Infor-
    mation Office. There is a table of contents with various sites on
    geoindicators, one of which is soil erosion. The site is broken into sev-
    eral sections. The first section is a brief description of erosion: a de-
    tachment of particles of soil and surface elements through various
    fluvial processes and action of the wind. There are other sites to click
    on with more information: Sediment Geochemistry and Stratigraphy,
    Stream Sediment Storage and Load, and Wind Erosion. The next sec-

tion describes how erosion adversely affects plant life and decreases water supplies and organisms, such as algae, corals, and fish. The next section is titled Significance, which outlines the social and economic problems and how erosion is an essential factor to assess ecosystem health and function. There is a section on Human or Natural Causes, natural processes that are greatly increased by human activities, such as land clearance, agriculture, construction, and so forth. There is a site for soil quality for more information. Some of the other topics with brief descriptions are Environment Where Applicable, Types of Monitoring Sites, Spatial Scale, Methods of Measurement, Frequency of Measurement, Limitations of Data and Monitoring, Applications to Past and Future, and Possible Thresholds. At the bottom of the page are Key References and Related Issues/Assessment. In summary, this site is thorough and well organized and includes a lot of additional sites to obtain more information. The site is clear and relatively easy to understand and would be suitable for eighth grade and high school classes.

Title: Erosion, Weathering
URL: http://www.geography4kids.com
Grade Level: Grade 6 and up
Best Search Engine: www.google.com
Key Search Words: process of erosion
Review: Navigating the Site: At the home page of the Geography4Kids site, click on Land (the other choices are Energy, Sky, Earth, Water, and Climate). The menu down the left side of the screen at Land is as follows, top to bottom: Biosphere, Global Geometry, Ecology, Ecosystems, Biomes, Populations, Food Chains, Cycles, Soils, Erosion, and Weathering. When you click on Erosion, a menu of examples appears to the left of the text, including Flood Erosion, Hillside Erosion, Ocean Erosion, Rain Erosion, and River Erosion. The text portion includes color photographs, and important terms are shown in all capitals and bold (for example, denudation, weathering, mass wasting). At the bottom of the text section, you can click on Next Stop on Tour to proceed to the Weathering topic. The examples menu for Weathering includes Dune Formation, Frost Action, Hillside Erosion, and Landslide. If you want to view the examples for Erosion first rather than proceed to Weathering, you click on the menu of examples to link to the example you want. The examples include one or two color photographs to illustrate the particular topic, accompanied by a one paragraph explanation. Once you have viewed all of the ex-

amples for Erosion, you have to return to Erosion in order to proceed to Weathering. Furthermore, it is not clear when you click on Next Stop on Tour what that Stop will be; sometimes it is the next item on the menu, sometimes it skips a couple of items. Given these two situations, I find the site navigation a bit confusing. I took a look at the site map, but it didn't clear up my confusion.

Site Review: The entire Geography4Kids site is colorful, playful, and graphically appealing. The text is written in a breezy style that would appeal to younger students (i.e., middle school). While meant to be fun, the presentation does not dilute the accuracy or substance of the material. There is a considerable amount of information covered which would be equally useful to adults looking for an introduction to the subjects included. The pictures used to clarify the text are exceptionally well chosen. The issue I raised earlier about site navigation is relatively minor. What I especially like about this site (and the companion sites Biology4Kids, Chemistry4Kids, and Physics4Kids) is twofold: it is designed to reach its target audience—kids—and the sense of fun makes the subject matter—science—engaging for them.

Title: Erosion—An Encyclopedia Article Titled "Erosion"—MSN Encarta Encyclopedia
URL: http://encarta.msn.com/encnet/features/reference.aspx
Grade Level: Grade 6 and up
Best Search Engine: www.metacrawler.com
Key Search Word: erosion
Review: This page is part of the MSN Encarta Encyclopedia site, which contains listings on just about every subject. The page has a clean design, but unfortunately there are several advertisements and other distractions located throughout the page. The page begins with a definition and general information on erosion; this area is concise yet complete in its information. The page then scrolls down into subtopics discussing the types of erosion and the effects of each, such as wind erosion and coastal erosion. Throughout these sections there are text links that take the user to related subjects or topics that dig deeper and further explain the information being reviewed. There are also "print this section" text links that allow the user to cleanly and easily print out the information—a very nice feature for anyone conducting research. The information concludes with a section focusing on the Human Impacts on Erosion, which is a relevant and important addition to the information given and would likely be welcomed by anyone seeking information on this topic. Overall this is a great page

that provides a comprehensive look at the topic of erosion, as well as related subtopics. I think it would be an excellent source for students seeking general information on the subject and especially for those students preparing a research paper or position paper. My search did not return a lot of sites that were dedicated to the topic of erosion in a general, explanatory sense—more sites that focused on products for dealing with erosion. Therefore, even though this is part of an encyclopedia site and the information is presented in that fashion, it is a great resource that provides excellent information on the topic.

Title: Erosion
URL: http://www.ttsd.k12.or.us/District/curriculum/elem/iscience/
     erosion.html
Grade Level: Middle school and high school
Best Search Engine: www.dogpile.com
Key Search Word: erosion
Review: This Web site lists various amounts of additional Web sites useful for both students and teachers on the topic of erosion. Some of the Web sites are suitable for just teachers, others are suitable for just students, while others are suitable for both teachers and students. The sites suitable for teachers are marked with a * next to it and the sites suitable for students are marked with a %. The Grand Canyon, the Iguazu Falls, the Oregon Coast, soil erosion, and water erosion are just some of the topics that this site links to. Out of all of these additional sites, one site on erosion in general stuck out like a sore thumb, supplying the reader with incredible information on erosion. This information is presented to the reader through a slide show containing 34 slides on erosion. This slide show is available at http://ucs.byu.edu/ bioag/aghort/282pres/Erosion/sld001.htm. Not only is information on erosion available, but clear, precise photographs representing various types of erosion such as soil erosion and water erosion are also available. These photographs illustrate and make more clear the information provided in the slide show on erosion. All in all, this Web site demonstrates a different way to provide a lot of helpful information to a reader. The information is presented in a way that both students on the middle school and high school level will understand.

Title: Demonstration Erosion Control
URL: http://www.colostate.edu/Orgs/CRSS
Grade Level: High school and above
Best Search Engine: www.yahoo.com

Key Search Word: erosion

Review: This Web site is very well designed. At the top of the site under the title is a large picture of some area where erosion took place. Every second that picture changes to another area. To the left of the picture is a section where you can click on circles to find more information. Included in separate circles are the titles, about Demonstration Erosion Control (DEC), Benefits, Sample Sites, Projects, and agencies. Under the picture is an area that describes what the DEC does. Following that are different links that you can click on to find more on the DEC. These links include: Read more about the DEC project, Learn what benefits DEC offers, Visit example sites, View a List of Projects by state, and Visit the Contributing Agencies. I liked this Web site because of the changing photos.

Title: Saving a Scenic Drive
URL: http://www.erosion.com
Grade Level: Grade 5 and up
Best Search Engine: www.metacrawler.com
Key Search Words: erosion + beach
Review: This Web site begins with two pictures that showed a beach before treatment and 10 months after treatment. It was amazing how much more beach was there after treatment. The Web site explained the breakthrough in beach restoration technology and how it reverses erosion without the addition of artificial fill and with no adverse side effects to adjacent shorelines. It explains how this system has restored more than 100 miles of beaches on the Great Lakes and ocean shorelines. On the left of the page you can click on recent projects, and it will show you recent projects that the program is working on. There is also a section on slide shows. If you click on that it will give you information and before-and-after pictures of beaches that have erosion. This Web site is out of Florida. It has many ways that you can contact the organization to get more information on this topic. There is also another Web site that you can go to.

Title: Science Lesson Plans
URL: http://www.col-ed.org/cur/science.html
Grade Level: K–5 teachers
Best Search Engine: www.metacrawler.com
Key Search Words: erosion, elementary
Review: At the home page, there is a list of various lesson plans. I clicked on the one for soil erosion and was given an overview of the subject,

along with lessons for kindergarten through fourth grades. The plans included the purpose, objective, activities, and materials needed. I was able to get lessons dealing with landslides, glacial erosion, water erosion, and sand erosion. At the end was a summary of all these topics. This is a great place for soon-to-be new teachers like me to steal some easy and fun ideas.

Floods

Title: Flash Flood!
URL: www.weathereye.kgan.com/cadet/flood/about.html
Grade Level: Grade 6 and up
Best Search Engine: www.google.com
Key Search Words: teaching, floods
Review: This is a terrific site for any teacher interested in finding an introductory lesson on floods. The home page provides a brief description of what floods are, as well as various links to stories of recent areas damaged by floods this year. Located at the top of this home page is a menu of different areas within this site helpful for those teaching students about floods and natural disasters. The first option, Flood Types, explains the types of floods, which include coastal floods and flash floods. Flood Prep prepares you for a flood with links to explain and help you prepare disaster plans and survival kits. Safety Rules is where you will find out what you should and should not do during a flash flood. Flood Quiz is a review of the information in this lesson as an online quiz, giving you instant results. Flood Watch will let you search the Web for flash flood warnings, allowing you to record your own findings on a weather map. Lastly, The Teacher's Guide is for teachers who want to read about the lesson and plan objectives for the Flash Flood lesson.

Title: FEMA for Kids
URL: http://www.fema.gov/kids/index.htm
Grade Level: Grade school
Best Search Engine: www.askjeeves.com
Key Search Words: what is a flood
Review: This site was developed by the U.S. Federal Emergency Management Agency (FEMA) to educate children on what to do in the event of a natural disaster. The home page is bright yellow and briefly describes the purpose of FEMA. At the bottom the page there are 10 links. When the "search" link is clicked, nine links are labeled down

the center of the page. At the bottom of this page is a keyword search prompt. When the word "floods" is entered into the keyword search, 84 links are provided. The first link is called "FEMA for kids: floods." When this link is clicked, the sound of rushing water welcomes you to the page. A paragraph that explains floods in easy terms is provided. Then the definition for Flood Watch or Flash Flood Watch, Flood Warning, and Flash Flood Warning are provided. In large print at the bottom of the page, a link to "The River Rises: The Disaster Twins Flood Story" is provided. On the right a large green button is shown. In this button are two links: Flood Disaster Math and Water, Wind and Earth Games. When "The River Rises" link is clicked, a story about the Disaster Twins visiting their grandmother is provided. The reader clicks the pages of the story, which tells of the family's experience when a flood warning is issued. Each page has icons that can be moved by the reader. These icons are examples of important themes from the story. For example, when the family had to pack their belongings to go to the Red Cross shelter, some of the icons depicted were a suitcase, a flashlight, batteries, and bottled water. When Flood Disaster Math is clicked, there are seven multiple-choice math questions available. Students can answer the questions and discover what their score is at the end. When Water, Wind and Earth Game is clicked, a game similar to Rock, Paper, Scissors is presented. The reader can choose the icon for water, wind, or earth. The computer tells the player whether or not they won the match. The score is shown on the bottom of the page. This Web page is appropriate for children in or above second or third grade. Most of the pages have brief textual information. Pictures of various natural disasters can be found on the search link. This Web page provides lots of information about different natural disasters and what children should do if they are ever faced with these events. The site also has interesting games and educational experiences for children.

Title: Significant Floods in the United States during the 20th Century–
USGS Measures a Century of Floods
URL: http://ks.water.usgs.gov/Kansas/pubs/fact-sheet/fs.024-00.html
Grade Level: Middle school and up
Best Search Engine: www.metacrawler.com
Key Search Word: floods
Review: This Web site is the "place to be" if you want information on any flood that took place in the United States during the 20th century. There are pictures, charts, descriptions, damage reports, measurements, and much more about each and every flood. The Web site be-

gins with a disturbing picture of the flood in 1993 in Mississippi. Then, there is a table of contents in which the user can click on the desired item. There is a section that will teach you how the USGS takes measurements of floods and how it categorizes them. The pictures of the floods are amazing overhead shots from helicopters. There are many graphs and charts that organize the most significant floods of the 20th century. There is a section that explains how the USGS gauges flood warnings for flash floods. Also, this site explains what a driver of a vehicle should do if he or she has to travel through a flood (with an animation). I give this site an "A" and I consider it one of my favorite sites.

Title: NOVA Online Flood!
URL: http://www.pbs.org/wgbh/nova/flood
Grade Level: Grade 6 and up
Best Search Engine: www.yahoo.com
Key Search Word: floods
Review: This site is the companion site for the PBS *Nova* broadcast titled "Floods!" The opening page gives a content listing for the topics covered in the program and on the Web site along with the choice of reading a complete transcript of the program. The topic list includes "Dealing with the Deluge," "Reconstructing a Floodfight," and a page with resources and a separate Teacher's Guide for use with the program (which can be purchased from *Nova*'s online store). The site has a lot of information about flooding, keying into three examples of rivers where humans have intervened to reclaim the rich land from the flood plain. There is a good overview of the program, highlighting the key topics covered. The availability of a transcript is a great resource for more information, and there is a resource page on the Web site for more information. This is an easy site to navigate. It has good information and interesting photos that illustrate the text. It is written in an easy style for a wide audience.

Title: Weather Safety: Floods
URL: http://www.weather.com/safeside/flood
Grade Level: Grade 2 and up
Best Search Engine: www.excite.com
Key Search Words: floods education
Review: This is a section of weather.com's Web site that teaches you everything you would ever want to know about floods. The first page contains definitions for three types of floods. A river flood is defined

as a high flow or overflow of water from a river or similar body of water, occurring over a period of time too long to be considered a flash flood. A flash flood is defined as a quick rising flood occurring from the result of heavy rains over a short period of time. The third type of flood is coastal flooding, which occurs when strong onshore winds push water from an ocean, bay, or inlet onto land. Other sections of this site include: Are You at Risk?, Before It Occurs, During a Flood, After the Flood, Flood & Flash Flood Facts, and Flood Index. This site offers a lot of good advice about what to do about flooding. I've lived on the beach my entire life and have been the victim of many a coastal flood, and there's stuff on this site that I didn't know and probably should have. The section on flood facts is pretty interesting too. I highly recommend a visit to this site if you're at risk for flooding in your neighborhood.

Fossils & Fossil Record

Title: Welcome to Fossils of New Jersey
URL: http://home.earthlink.net/~skurth/
Grade Level: High school
Best Search Engine: www.yahoo.com
Key Search Word: fossils
Review: This was a very interesting Web site to look at. I like it very much since there are many interesting pictures. The main part of this site is of pictures of types of fossils found in New Jersey. When you look at each picture you are capable of hitting the title of the fossil. Once you do that you are directed to information on the specific fossil. This site also gives you a local organization link at the bottom of the page. The last section of the site has links that you can use to find more information. These links include: Hadrosaurus foulkii—World's first dinosaur skeleton, North Jersey Paleoworld, Academy of Natural Science of Philadelphia, American Museum of Natural History, Beach fossils from NJ and NY, and NJ State Museum.

Title: Getting into the Fossil Record
URL: http://www.ucmp.berkeley.edu/education/explorations/tours/fossil
Grade Level: Middle and high school
Best Search Engine: www.dogpile.com
Key Search Words: fossils and fossil records
Review: This Web site is sponsored by the National Science Foundation and is an excellent site for students to better understand fossils and

fossil records. This site supplies the reader with an animated demonstration and tour showing how a dinosaur could get buried by sediment after it dies, become a fossil, and then become exposed. In the animation tour the students will learn what fossils and fossil records are, how organisms become fossils, why many fossils will never be found, and where paleontologists look to find fossils. As one works their way through the animated tour, the green side bar on the left side of the Web site shows the reader the topics they have covered. At any point throughout the tour, the reader has the opportunity to click on any of the underlined dark red words to see a definition of that word. This will help the reader better understand the written information provided throughout the animated tour. In addition the animated tour asks the reader to answer some questions related to fossils and fossil records in order to move on throughout the rest of the tour. This keeps the reader on his or her toes. All in all, this is an excellent and fun way for students on both the middle school and high school levels to learn about fossils and fossil.

Title: Proteacher
URL: http://www.proteacher.com/110055.html
Grade Level: Grade K–5
Best Search Engine: www.metacrawler.com
Key Search Words: fossils, elementary
Review: This Web site gives you everything you need to know about fossils and dinosaurs with a host of lesson plans, recipes, songs, pictures, and related links. I clicked on "dinosaur babies" on the home page and got paleontology lesson plans for grades 1–6, with kids learning about real fossils and then making their own edible fossils. Great site for teachers to get new and creative ideas!

Title: Extinctions Inc.
URL: www.extinctions.com
Grade Level: Adults
Best Search Engine: www.metacrawler.com
Key Search Words: fossil + fossil records
Review: This Web site has links to fossil stores as well as fossil showcases. Links to the information pages and other features, such as a fossil show schedule and fossil news, are located on the left of the page. Fossilsforsale.com is a massive new fossil store with a lot of fossils online. If you click on fossil news, you can get information that is new and information going back all the way to 1997. Extinctions Inc. is a

business dedicated to supplying fossils to museums and collectors around the world. This Web site was not full of information, but had fossils for sale. I found this Web site to not be very helpful. The reason I reviewed this Web site was that I found it very annoying that it stated that it had fossil information and fossil records but really they just wanted to sell their fossils. I came across more sites that I found were more helpful, such as www.fossilrecord.com

Title: Fossils! Behind the Scenes at the (Royal Ontario) Museum
URL: http://www.rom.on.ca/quiz/fossil/
Grade Level: Sixth grade and up (and younger proficient readers)
Best Search Engine: www.metacrawler.com
Key Search Word: fossils
Review: Navigating the Site: At the main menu of the Fossils! site, the following choices appear to the right of corresponding color graphics and photos: "To learn how fossils are formed and who studies them—click here!" "To discover how and why fossils are prepared—click here!" "To see how and where fossils are collected—click here!" "To find out how scientists identify and classify fossils—click here!" "To play a game with fossils—click here!" "To learn more about the ROM's fossil collection—click here!" Keywords in each menu selection are in green. The word "here" provides a link to the topic. Key terms in each topic are in boldface type. At the end of the first topic, there is a link to an introduction to the Royal Ontario Museum's Department of Paleobiology, which contains some introductory material to both fossils and to the department, as well as links to collections, research, exhibits, and other Paleobiology links. At the end of each topic, a link from the word Fossils! returns you to the Fossils! main menu.
Site Review: The Fossils! site is appealing from a graphics design perspective. It is clearly intended for younger (middle school) students. Each topic is about one page long, with short, easy-to-read paragraphs, wonderful color graphics, and plenty of white space. The language is straightforward, written to appeal to children. The site contains basic information, best suited to the level of younger students, but fine as an introduction for anyone. I loved the section "So How Does an Animal Become a Fossil?" It was cleverly illustrated, colorful, easy to understand, and memorable. I especially liked the fossil game. It was challenging to try to match some of the fossils with the living animals. The game is a fun learning tool for children, with instant feedback. I really like finding sites like this one, which are designed for younger students and strike a good balance between enjoyment and intellect.

Title: The Fossil Record
URL: http://www.es-designs.com/geol105/Topics/fossils.html
Grade Level: College freshman
Best Search Engine: www.metacrawler.com
Key Search Words: fossil and fossil records
Review: This Web site is a combination of lecture notes from a college
    course, Geology 105: History of Life, and additional information on
    related topics. The home page, titled The Fossil Record, is an
    overview of how fossil records are a primary source of information for
    evaluating the history of life. There are lecture notes on fossils and a
    study guide. The notes are an outline that include the following: def-
    inition of fossils—remains or traces of once living organisms, charac-
    teristics—hard parts, soft parts, traces; the history of fossil use/early
    interpretations—collected since prehistoric times (that is, formed in
    rocks through celestial influences, creations of the creator); correlat-
    ing rock units, building a relative time scale and historical facts of fos-
    sil records. The outline also includes biases in records due to physical,
    ecological and human influences, plus the limitations of fossil
    records. In addition to the outline, there are topics listed on the left
    side of the home page on these subjects: Geology and Time—how
    fossil records are recorded; Evolution and Diversity—important bio-
    logic processes in earth history; Origins of Life and Earth—the begin-
    ning of the universe, solar system, the development of cells. This
    site is user friendly, thorough, and straightforward. The information
    would be suitable to an introductory college class.

Title: Dinosaur Fossils
URL: http://www.enchantedlearning.com/subjects/dinosaurs/
    dinofossils/Fossilhow.html
Grade Level: Pre-K through adult
Best Search Engine: www.google.com
Key Search Words: fossils and kids
Review: Wow, another fantastic site by www.enchantedlearning.com! At
    the top of the dinosaur fossil page you will find a colorful menu of
    links to many fossil topics. They are as follows, First Dino Fossil Dis-
    coveries, What are Fossils?, How do Fossils Form?, Types of Fossils,
    Finding Fossils, Dating Fossils, Excavating Fossils, Bony Jigsaw Puz-
    zles, and Famous Fossil Hunters. Underneath these you will find links
    to the various areas of the world where fossils have been located
    (North America, South America, Africa, Asia, Europe, Australia,
    and Antarctica). I chose to begin my adventure within the How do

Fossils Form? section. I found a great deal of information, including a flow chart on fossil formation. I was impressed! This section also offers six ways that organisms can turn into fossils. Next, I chose to enter into the Finding Fossils section. Here I learned that finding fossils requires a perfect balance between skill, tenacity, and luck. I also found out that children have contributed to this field quite a bit over the years. Next I clicked on a link to fossils found in the North American section of the world. Here the information can be organized by country and then even further by state. When I scrolled down to New Jersey I saw a list of six dinosaur fossils that have been found in our state. Enchanted Learning offers links to the state as well as to the dinosaur. These links take the user to even more information about the subject. I did not know New Jersey had a state dinosaur, the Hadrosaurus foulkii. The information on this site is infinite. One topic leads you to another, to another and so on and so on. Almost everything as a link to even more information it is almost overwhelming. I absolutely recommend this site to anyone wanting information on dinosaurs. All you need is the time to surf through the plethora of material. This site could be useful to anyone ages 4 through 94. www.enchantedlearning.com has done it again.

Title: Getting into the Fossil Record
URL: http://www.ucmp.berkeley.edu/education/explorations/tours/fossil
Grade Level: Section 1, Grades 5–8; Section 2, Grades 9–12
Best Search Engine: www.metacrawler.com
Key Search Words: fossils and fossil records
Review: This site is a perfect site for teachers and students seeking information on fossils and fossil records. This site is a link from a main page entitled Explorations Through Time, created and still under development by Pacific Bell, which features a number of links to other similar subjects related to dinosaur and fossils. The home page for fossils records, found at the address listed above, contains text and icon links to sections designed specifically for students and others designed for teachers. The teacher section gives terrific information, lesson plan ideas, and everything else needed to guide students through the use of the site and enhancing the experience while providing great ideas for activities and enrichment ideas. The site is designed for the teacher to work with the students in going through the interactive tour of information and questions on the subject. The pages designed for the students contain great animations to demonstrate how things become fossils and how fossil records are created, read, and used. There are also

text links found throughout the descriptive passages on the site that link to definitions of the relevant vocabulary words—a great feature. Some of the pages end with questions related to the information being covered that related to the information provided to the teacher in that section—another great feature. Most of the questions are interactive and provide text links so that a correct answer will link to a further explanation of the answer the student selects, but an incorrect answer leads them back to try again to ensure they receive the correct information and description. There are also links to slide shows that dig deeper into some of the diagrams and information covered in the site that allow the students to see exactly what is being described. Overall, I consider this site to be outstanding for teachers and students. The site is so well done and organized that it is unlikely anyone seeking information on this subject would be disappointed. It is a colorful, well-designed and highly interactive site that would keep anyone's interest. The activities and animations are well selected and highly effective in teaching about fossils and the fossil record and the information is so comprehensive in every aspect, right down to the smallest of details. I think this is one of the best sites I have reviewed for students, as well as teachers and would love to use it in my own classroom someday if I am teaching about this subject.

Geochemistry

Title: Organic Geochemistry Research Group
URL: http://ks.water.usgs.gov/Kansas/reslab
Grade Level: Grade 12 and up
Best Search Engine: www.yahoo.com
Key Search Word: geochemistry
Review: The site comes to us from the University of Kansas with a strong collaboration with the U.S. Geological Survey office. This is a well-laid-out site, nicely organized with a lot of information. It is not for the faint of heart. This site is for the serious professional or lay person/ student looking for information about clean and polluted waters. The opening sentence on the main page really tells the story about what will be found in this site: "The mission of the Organic Geochemistry Research Group is to understand the organic geochemistry of natural and polluted waters." They are serious. There appears to be an extensive resource list of articles on one page, a large bibliography on another. This site is packed with information about the geochemistry of water and the ongoing research efforts. The professional could use this

site fairly extensively for a historical perspective. There is a page labeled Education and Other Information. There are some good ideas, but not a lot of in-depth lesson plans. But it's a good starting place.

Title: Geochemistry Department
URL: http://www.aist.go.jp/GSJ/dGC/chemtop.htm
Grade Level: Grade 8 and up
Best Search Engine: www.excite.com
Key Search Words: geochemistry education
Review: This is about the best I could find for geochemistry. This site is the home page of the geochemistry department of the geological survey of Japan. The site explains that they are the department responsible for geochemical studies in relation to many geological research projects. They research the behavior and distribution of major and trace elements and isotopes in rocks and minerals including extraterrestrial material. There is a colorful map that is the geochemical atlas for the North Kanto area of Japan. If you click on the link below it you can get to different geochemical maps of the world. The only other section offered on the link is a section titled "analytical methods," which explains how the department gets its data, what machines and methods the researchers use, and the names of the people on the research team. Overall, this was a very poor site for geochemistry. If you're interested in becoming a geochemist, you may like to read about some of the methods of research they use. But for the passerby like myself, I didn't enjoy the site very much.

Title: History of Geochemistry
URL: http://www.smarterscience.com/geochis/geochis.html
Grade Level: High school and above
Best Search Engine: www.yahoo.com
Key Search Word: geochemistry
Review: This page is divided into seven links that allow you to jump to topics of interest on the page. These links are listed at the top of the page: Aqueous and environmental geochemistry, Geology and geochemistry, Impact of physical chemistry, The phase rule and the Phase Ruler, Tying the knot between the two disciplines, The Geophysical Laboratory and Geochemistry on its way to the 21st century. These seven links are organized to give a detailed history of geochemistry. This page has copies of original source material that provides proof of the origins of geochemistry. It tells how geochemistry evolved into the science it is today and discusses possible implica-

tions of the progression of the science in the future. Navigating this page is easy. The seven main links at the top of the page allow the reader to jump to an area of interest. If the reader wants to read the entire history, they can scroll down the page. The pictures provide interesting detail to the textual information provided. This site could be used by teachers as a tool to link science and history. It is an excellent site for anyone interested in the history of science and provides interesting pictures and primary historical sources.

Geodata

Title: Ambon Information Website
URL: http://www.websitesrcg.com/ambon/
Grade Level: College
Best Search Engine: www.yahoo.com
Key Search Word: geodata
Review: This was a very hard Web site to look at and to follow. There seemed to be so much information on the page that it is confusing to the viewer's eyes. I definitely wouldn't recommend the site. Included in the Web site is: a picture of a map on the top left of the section. To the right of that area is a section where new information had been posted. Under that section is information on different countries. Following that section was articles.

Title: About Geography
URL: http://geography.about.com
Grade Level: Grade 7 to adult
Best Search Engine: www.google.com
Key Search Words: geodata, geographic data, geography, GIS
Review: Once I figured out that the term "geodata" referred to geographic data, I found this site, which is intended to educate students about geography. I then set out to try to educate myself about geo(graphic)data and what it is used for. Navigating the Site: At the left side of the home page is the following menu of topics covered in the site: Basics About Geography; Blank/Outline Maps; Cartography; Census/Population; Cities & Transport; Climate & Weather; Clip Art; Country Facts; Cultural Geography; Disasters/Hazards; Finding Places; Fun, Games & Humor; Geo Education; GIS & GPS; Historic Maps; Homework; Help; Large Cities; Latitude/Longitude; Maps; Photos; Physical Geography; Rivers and Streams; Street & Road Maps; Time & Time Zones; Topographic Maps; U.S. Maps; World Maps; World Population; ZIP

Codes; and Subject Library. I familiarized myself with the site by researching the topic I was trying to understand. About Geography contains links to topics including Glossaries and Dictionaries, which
includes among many choices both a site glossary and a link to the
About Geography Information GIS Dictionary (described as definitive
for GIS terms). I used both sources to look up "geographic data" and
"GIS," which by this point, I had discovered, used geodata. The Geography Basics menu selection led me to All About Geography, an article
that contained a section titled "What is the Future of Geography?"
which in turn contained a link to "GIS (Geographic Information Systems)". This led to a page full of links to articles about GIS, both on the
About Geography site and on other sites. I looked at "An Introduction
to GIS" and "An Interview with a GIS Specialist" on About Geography, and at "Geographic Information Systems" on the USGS (U.S.
Geological Survey) site. I also selected GIS & GPS from the main
menu, which leads to a page titled "GIS, GPS and Technology in Geography" and contains links to other sites. From this page I linked to
the GIS Lounge site, which was noted as "Best of the Net." I found a
great article, "What is GIS?"; it is fairly easy to read (but probably more
suited to high school students) and contained an interesting section on
the data used by a GIS, referred to as a geodatabase. As is obvious from
the site menu shown above, there is a lot more information on geography in this site.

Site Review: I liked this site because it is meant for students and it is loaded
with information. It was not a graphically pleasing site in that it does
not contain color graphics, nor does it use graphics to make the site appealing; it is strictly textual information. The site is easy to navigate
from the main menu, which is comprehensive. Multiple menu choices
contain overlapping information, so that there are multiple routes to
the data you are trying to find. The language used in the articles is appropriately technical and highly readable. What I do not like about this
site is the degree of commercial advertising it contains (the only color
graphics on the site!), which I find both annoying and inappropriate
on sites designed to be used by students. What I especially liked about
this site is that it is comprehensive enough to allow me to find information on what seemed to be an obscure topic.

Title: Geodata
URL: http://www_ai.ijs.si/~ilpnetz/apps/geo.html
Grade Level: Teachers of all grades
Best Search Engine: www.metacrawler.com

Key Search Words: geodata, elementary

Review: This was the hardest site for me to locate . . . I checked other search engines as well but came up empty. The data base I finally found contains a simplified description of rivers, roads, railroads, woods, and buildings. It gives at least one fact per subject and it corresponds to real world topographical data. Each object is discussed by its geometry using a two-dimensional projection. It uses basic concepts that one would see on a map. All I can tell you is that students wouldn't be able to maneuver this site because I had difficulty myself. Do yourself a big favor and skip this one!

Title: GeoData Institute
URL: http://www.geodata.soton.ac
Grade Level: High school
Best Search Engine: www.metacrawler.com
Key Search Word: geodata

Review: When you first open to this page you have a choice to go to Environmental Services, Multimedia Software Development, or GIS Consultancy & Services. On the left of this page you can click on About Us, Find Us, Contact Us, Publications, and Hosted Sites. I clicked on the Environmental Services. When you click on this you have the choice to click on Environmental Services, Projects, and Links. I clicked on Environmental Services. When you click on that you have a choice of the main environmental services that they provide, such as Consultancy, Strategic Planning, and Information Management. I clicked on Information Management. GeoData provides a wide range of information management services to support environmental applications. I did not find this Web site to be very helpful or descriptive about geodata. I am not sure if this Web site was at all what you were looking for, but it is helpful to know that this search engine can pull up information that is not even what you are looking for. This site may be helpful if you are looking for information related to the environment and geodata.

Title: New Jersey Geological Survey Digital Geodata Archive
URL: http://www.state.nj.us/dep/njgs/geodata
Grade Level: Grade 9 and up
Best Search Engine: www.yahoo.com
Key Search Word: geodata

Review: This site is hosted and maintained by the New Jersey Department of Environmental Protection and the state of New Jersey. This particular site focuses on geological information and has a number of subtopics and links for the visitor to explore. The navigation bar to the left provides several main links that when clicked on take the user to more information on that topic. The two main links/areas I used were Geodata and Education. Geodata takes the user to a page that provides a series of links for four main topics: Base Maps, Geology, Groundwater, and Geophysics. When a text link on a particular topic is clicked on, the new page is an article or abstract detailing information on the subject. The Education link takes the user to a page that offers products for sale, free online resources, lesson plans, and info circulars. Similar to the rest of the site, the links are all to areas that contain text and downloadable information. The site is one of a few that I was able to find on the subject. None of the ones I selected from my search were really geared toward education. I selected this site because it is a New Jersey site that did in fact contain information on the subject and had useful tools for teachers. Students could use the site to retrieve information, but would likely require the assistance of a teacher in further understanding the subject. For those who have an understanding of the subject, the site would be useful in that it offers some tools, maps, and other animations that explore the topic. There are some links listed that may be helpful to students and teachers in further exploring the topic as well. Overall, I think the site is a solid resource, but mostly for those with an understanding of the subject and in higher grades.

Title: Thurston Geodata Center
URL: http://www.geodata.org
Grade Level: Honors high school students
Best Search Engine: www.dogpile.com
Key Search Word: geodata
Review: This Web site is maintained by the Thurston Geodata Center (TGC), located in Washington state. The Web site provides support and services to federal, state, and local agencies, private businesses, and the general public. The site also provides Thurston County staff with accurate spatial geographic information and provides access and support in using information in their daily operations. One of the great things about this Web site, because it is somewhat difficult to understand and surf through, is that the site provides a site tour to the visitor. This tour is very helpful and makes the Web site easier to understand. One of the main aspects of the Web site is that one can create his or her own map. Creating this map requires a lot of detail and

can be difficult to do, but with time, it can be completed. Additional aspects to this Web site are that data sources are available for viewing and/or purchase, information about TGC is available, TGC staff information is also available, and there's a coming-soon section. All in all, this Web site may be one of the more difficult sites to surf through, but there is some good and helpful information available to students. The students just may have to put in a little extra time to locate this information.

Title: A Comprehensive Data Model for Distributed, Heterogeneous Geographic Information
URL: http://www.regis.berkeley.edu/gardels/geomodel
Grade Level: College sophomore–senior
Best Search Engine: www.metacrawler.com
Key Search Words: types of geodata sources
Review: This site has an extensive table of contents. This review will cover some of the key points of the site. The first section is the Meaning of Geographic Information, which comprises data about surface, subsurface and the atmosphere of the Earth, interpretations of data and a framework to understand the information. The next section is Mapping and Spatial Thinking, or how we use the data. There is information on Realms of Geodata, for example, an Earth model describing continents, landforms, landscape, and environment. Other sections are Geospatial, System Models, Language of Geodata (with diagrams), Geodata Modeling (coded or computerized representation or abstraction of real-world entities and phenomena), Geodata Structures (with diagrams), maps, Sharing Geodata, Conclusions, and References. This site is produced by the University of Berkeley and is quite technical. There is a lot of information and each section is well written and concise. This site is suitable for college level, most likely sophomore through senior year.

Geographic Information Systems

Title: Geographic Information Systems
URL: www.usgs.gov
Grade Level: Adults
Best Search Engine: www.askjeeves.com
Key Word Search: what is geographic information systems?
Review: The top of the page gives a brief example of what types of activities GIS can be used for. Then the site lists four main questions that it

answers with either a brief description given on the page or a host of different links under the question. The questions are: "What is GIS?," "How does GIS work?," "What's special about a GIS?," and "Applications of GIS." Under "What is GIS?" a brief definition of GIS is given. Under "How does GIS work?" the following six links are given: Relating Information from Different Sources, Data Capture, Data Integration, Projection and Registration, Data Structures, and Data Modeling. These links give textual and graphical information for each topic. The subheadings for the last two questions are formatted in the same manner. The site is organized in an outline fashion, which makes it very easy to navigate. The information is clear and concise. This site is a good resource for anyone interested in GIS and the various data that can be extrapolated from it. While this site is rather boring to look at, it displays information in a user-friendly manner.

Title: Geographic Information Systems
URL: http://www.us.gov/research/gis/title.html
Grade Level: All grade levels
Best Search Engine: www.google.com
Key Search Words: geographic information systems
Review: The main page tells you all about GIS. When you scroll down you can find out different information. The first option is: How does GIS work? Under the title there are different links: Relating Information from different information sources, Data Capture, Data Integration, Projection and Registration, Data Structures, and Data Modeling. When you double-click on any one of the above, you are able to view and enlarge different pictures and graphs. Scroll down on the main page: What Is Special about GIS? Here are your options to explore: Information retrieval, Topological Modeling, Networks, Overlay and Data Output. Back on the main page: scroll down to Application of GIS. Here are your options: GIS through History, Mapmaking (fun), Site Selection, Emergency Response Planning, Simulating Environmental Effects, Graphic Display Techniques, and The Future of GIS. At the bottom of the page you can access the USGS home page. This site is great for students who are new to Geographic Information Systems. This site is meant for beginners.

Title: The Geographic Information Systems FAQ!
URL: http://www.census.gov/ftp/pub/geo/www/faq-index.html
Grade Level: Grade 10 and up
Best Search Engine: www.yahoo.com

Key Search Words: geographic information systems

Review: Talk about no nonsense—here is everything (and then some) you wanted to know about Geographic Information Systems. The list of questions is nicely laid out, neatly indexed by category. The answers to the questions are well written, to the point but understandable. Citations are used throughout to provide additional resources and the sources to the answers. While the site uses information from the U.S. Census bureau, there is a disclaimer at the bottom distancing the bureau as a participant in the site. This is an all-inclusive site offering information and other resources about GISs. I have noted this as 10th grade and up, but any student who is into science at a younger age could find this interesting. It is not really for the casual viewer. Some of the questions are asked in a slightly tongue-in-cheek manner but the answers are serious—information only and to the point.

Geological Time Scale

Title: Take Our Web Geologic Time Machine
URL: http://www.ucmp.berkeley.edu/help/timeform.html
Grade Level: Middle school and up, students and teachers
Best Search Engine: www.metacrawler.com
Key Search Words: geologic time scale
Review: This Web site gives you all of the different eons and times and eras within them. Within these eras, it breaks down time even more. For instance, within the Pleistocene eon is the Cenozoic era and within that are the Quaternary and Tertiary periods. Within the Quaternary is the Holocene time. Anyway, I chose Pleistocene and got info on temperature, plants, insects, birds, and mammals of its day, complete with pictures. I was able to click on other links found as well. The site is not for the very young but would be great for middle school students on up. Great for doing reports!

Title: Geologic Time Scale
URL: http://www.enchantedlearning.com/subjects/Geologictime.html
Grade Level: Elementary through high school
Best Search Engine: www.dogpile.com
Key Search Words: geologic time scale
Review: The Geologic Time Scale section of the Enchanted Learning site begins with an interactive color graphic depicting continental drift with a link to a terrific article on continental drift and plate tectonics

elsewhere on Enchanted Learning with effective interactive color graphics. This is an excellent Web site for students of all ages for gathering helpful information on geologic time scales. It is a colorful, eye-catching, and fun site to surf through. At the top of the page an explanation of a geologic time scale is provided. Under this explanation is an animated geologic time scale of the Earth and its land, showing the reader how Earth has changed throughout the years. The Web site allows the reader to see this animated geologic time scale going forward or backward in time. The animated geologic time scale is easy to understand partly because a colorful key is provided next to the animated diagram. The topic of the geologic time scale is comprehensively covered between the table's graphic summary and the multitude of links to more in-depth coverage of the eras, the epochs, and their pivotal events. Next, within the Web site is a detailed, colorful, and informative geologic time scale starting with the Cenozoic Era and going all the way up until today. Throughout this scale, the reader can click on many words to find additional information about a particular word. Pictures of specific animals are provided next to the time periods with which the animal lived in. By doing this, the Web site gives the reader a mental image of when that particular time period took place and what things were like. Other links take you to articles on the University of California, Berkeley, Museum of Paleontology (UCMP) site. The vocabulary of this topic is difficult, and the links to the UCMP site lead to a more difficult reading level. All in all, this is an excellent Web site for students of all ages to use when gathering information on geologic time scales. It is an easy and fun site to surf through.

Title: Geological Time Scale
URL: http://www.geo.ucalgary.ca
Grade Level: High school
Best Search Engine: www.metacrawler.com
Key Search Words: geological time scale + diagram
Review: When you first open this page it is overwhelming with the text. This Web site is mostly information. It does not contain pictures or graphs; it just shows the diagram of the time scale. This Web site begins by explaining that geological time is often discussed in two forms. Those two forms are relative time and absolute time. The site describes each of these and gives examples that are very helpful and easy to understand. The time scale is displayed with the oldest at the bottom and the youngest at the top. The present day is at the zero

mark. I did not really know much about the geological time scale but after reviewing and reading this Web site I felt a little better about the topic. I do however feel that you would need to go to other Web sites that have more information on this topic. At the bottom of this Web site you can click on a button that will take you to the Geological home page. I went to that site and I found it to be full of information, links, graphs and diagrams. It was a great source to search and find information related to geology.

Title: Basics of Geologic Time
URL: http://geology.about.com/cs/basics
Grade Level: High school senior–college freshman
Best Search Engine: www.metacrawler.com
Key Search Words: geologic time scale
Review: This site reviews the basic resources about geologic time and how we track it. The first section—How Old Is the Earth? Why Should We Care?— a commentary by a professor at the University of Oregon. Another section, Mapping Deep Time, looks at how we navigate the span of time in Earth history. There are Earth History Maps with graphics and a Geologic Time Machine with charts of each era. There is a section on Pseudonumerology, or how to memorize the geologic time units. There is a set of links: Official Geologic Time Scale, Measuring Deep Time, and Dating Methods. This site covers technical information, but the language is fairly easy to interpret. The site has a detailed description of geologic time scales and how they are used to explain the history of time. This site would be suitable for high school seniors through college freshman.

Title: Geological Time and the Evolution of Earth
URL: http://www.owu.edu
Grade Level: Grades K–3
Best Search Engine: www.google.com
Key Search Words: elementary geology lessons
Review: This is a great Web site with ready-to-use geology lessons. These activities are appropriate for children in grades K–3, but they may be easily adapted to other grade levels. I think this is a great site because it allows the students to realize all of the events that took place on this planet before humans were around. The lessons are fun and for the most part done with everyday materials, the only exception being the fossils. Aside from geological time you can also find geology lessons on this Web site that are about earthquakes, plate tectonics, and

volcanoes. I think this site could be a valuable resource for any science teacher who would like to incorporate a geology lesson into the curriculum.

Geomorphology

Title: Geomofologia Wurtualna/Virtual Geomorphology
URL: http://hum.amu.edu.pl/~sgp/gw/gw.htm
Grade Level: College
Best Search Engine: www.yahoo.com
Key Search Word: geomorphology
Review: Obviously, I will only be reading the English version of this page, for I am unable to read Polish. The home page shows clickable images to choose your language preference (choice of two). The first thing I did was click on the university link. I found that it is a university for School of Polish Language for foreign students, located in Poland. From the university page there are seven links to the left, for the home page, school information, research, and so forth. There is also a link at the bottom for comments and suggestions. I chose to go back and click on the "introduction to virtual geomorphology" page. This is where the author writes about the project: what it is; what it does; why he thinks it's a good idea; whether or not people will help; and a section if you are interested in writing a section. On this page, he offers two additional links. The first is "world's drylands." The second is a link for authoring a section. (I checked the authoring link first—only because I know myself and will most likely get absorbed on the first link mentioned.) From the authoring link, instructions are given clearly on how to submit writing. There is also a reference link to mathematical symbols and how to communicate them via a keyboard. Oops! The drylands page doesn't work. (I use the rule of thumb: try it three times, then move on.) Moving back to the front page, where curiosity took over. There was a word "creationist" highlighted, so I clicked on that and it linked me back to the university, and then to a page entitled Evolution with a new Web page and a reminder to "change your bookmark." I am now led to a page called Science and Creationism (http://www7.national academies.org/evolution). This page has additional links on it. They are as follows: 1. Statements from the National Academy of Sciences. This link sent me to educational standards for Kansas as a joint statement from National Science Teachers Association, National Research Council, and the American Association for the Advancement of Science. It also had a link regarding teaching evolution in the classroom;

2. Statements from Other Science Organizations; 3. For Teachers; 4. Books and Videos; 5. Related Links and Resources; and 6. Suggest a Link or Resource. Curiosity took me completely off track. I am moving back to the front page. There are 13 links offered. As I went into the first link (geomorphological systems), under the table of contents, it escorted me to a page to select textbooks for geography courses. My opinion of this page: it is taking me many hours to still not have a quality definition of geomorphology and has accomplished making me feel less than adequately educated. I really do think that this page is good for those whom have scientific minds and/or teach the sciences. I found my answer to "what is geomorphology?" by using the Yahoo! Reference—Britannica Concise page. It gave me a clear definition that I could understand.

Title: Geomorphology Homepage
URL: http://erode.evsc.virginia.edu
Grade Level: Upper middle and high school
Best Search Engine: www.webcrawler.com
Key Search Word: geomorphology
Review: This site is maintained by Alan D. Howard, a professor in the Department of Environmental Science at the University of Virginia. It has incredible photos that will generate interest, ergo discussion, on the topic. Unfortunately, I could not find when it was last updated. This was the first time I explored the term "geomorphology" since 1974, and was able to get a feel for the topic by the photos.

Title: Geomorphology
URL: http://geoimages.berkeley.edu/geomorphology/wells/wells.htm
Best Search Engine: www.google.com
Key Search Word: geomorphology
Review: This site is a link from a site called Geo-Images Project that can be found at http://www.geoimages.berkeley.edu/geoimages.html. The Geo-Images Project is a way of using images, mostly pictures, to teach geography on the Internet. One of the page's many links goes to Images Illustrating Principles of Geomorphology by Professor Lisa Wells of Vanderbilt University. This site is basic, with many thumbnails that display different examples of geomorphology. The pictures are grouped into 12 categories.

1. Alluvial Plain Deposits and Human Transformations of Land Surfaces
2. Dunes

3. Coastal Zones and Volcanoes

4. Glaciers and Valleys

5. Desert Pavement/Varnish and Weathering

6. Flood Deposits

7. Coasts and Rain

8. More Coasts

9. Tidal Flats and Miscellaneous

10. Erosion and Rivers

11. Faults, Glaciers and Volcanoes

12. Miscellaneous Additions

While this site is not very detailed about what geomorphology is and the process of studying it, it does give several visual examples of what it is. A novice or a scientist looking for visual aids could find this Web site to be a great asset.

Title: Geomorphology from Space—A Global Overview of Regional
    Landforms
URL: http://daac.gsfc.nasa.gov/DAAC_DOCS/
    geomorphology/GEO_HOME_PAGE.html
Grade Level: Grade 9 to college
Search Engine: www.yahoo.com
Key Search Word: geomorphology
Review: This Web site provides an electronic version of a NASA publica-
    tion from 1986, which studied landforms and landscapes from space
    with photography and radar imaging. The Web site provides a multi-
    tude of educational information including text, pictures, maps, under-
    sea photography, and charts of information. Twelve subtopics include
    an introduction to geomorphology, tectonic geomorphology, volcanic,
    fluvial, deltaic, coastal, karst/lakes, eolian, glacial, planetary, mapping,
    and a discussion called Future Outlook. This site provides basic infor-
    mation about geomorphology in addition to discoveries made from
    space. The college level and advanced high school reader will be able
    to comprehend the information. This level of student will find the
    graphics interesting and helpful in the pursuit of understanding this
    topic and related issues. Navigation of this site was problem-free. Each
    picture provides icons that allow the printing of the pictures only. Most
    pictures were clear. The site includes Web connections to NASA and
    Goddard, and also e-mail addresses for the page author, the NASA of-
    ficial involved in the development of the page, and the Web curator.

Title: Glacial Geology and Geomorphology: a Journal of the British
    Geomorphological Research Group
URL: http://ggg.qub.ac.uk/
Grade Level: High school or college
Best Search Engine: www.metacrawler.com
Key Search Word: geomorphology
Review: This advanced-level British Web site is managed by editors W.
    Brian Whalley of the School of Geosciences at The Queen's University
    of Belfast and Martin J. Sharp of the Department of Geography at the
    University of Alberta at Edmonton. This site contains four areas: Pa-
    pers, Communications, Resources, and About GGG. The Papers sec-
    tion contains papers listed in order of publication with images. There
    are abstracts of the papers, which is very helpful. The Communications
    section contains editorials (though when I explored this area I received
    a "file not found" response) and a Discussion Forum. The Resources
    section contains links for related sites and Paper Journals, which con-
    tain tables of contents for some well-known journals. These also in-
    clude links to other sites. Lastly, the About GGG section contains
    information about the site, namely information about the editors and
    editorial staff. I would recommend this site for finding papers and jour-
    nal articles for research on the topic of geomorphology. I found the in-
    clusion of abstracts practical and time-saving for researchers and the
    inclusion of related links to be helpful also. It is a site I would recom-
    mend to teachers as useful for high school and college students.

Title: West's Geology Directory and Physical Geography
URL: http://www.soton.ac.uk
Grade Level: High school and college
Best Search Engine: www.metacrawler.com
Key Search Word: geomorphology
Review: Ian West is a professor at South Hampton University. The Web
    site includes an index of the Geomorphology section of West's Geol-
    ogy Directory. The 18 items can be easily selected or the individual
    can scroll down to find the items accompanied by a brief description
    and several related links. West's directory includes Beaches, Bodmin
    Moor UK, Caves, Cliffs, Dartmoor UK, Floods, General Geomor-
    phology, Geography-General, GIS, Glaciers, Karst, Landslides, Lists
    of Internet Links, Remote Sensing, Sea Level Changes, Tors, and
    Weathering. Navigating through the Web site can be challenging, if
    not frustrating, for a novice. Many links were either unavailable or
    had been changed or moved. Those sites that were accessible were

detailed. Many photographs, which could be enlarged and used for teaching purposes, were available. The site seems to be designed for those who are studying or working in related fields. This would not be the best selection for an introduction to geomorphology.

Title: Geomorphology
URL: http://encarta.msn.com/index/conciseindex/69/06913000.htm
Grade Level: High school
Best Search Engine: www.metacrawler.com
Key Search Word: geomorphology
Review: I found this article while searching through Metacrawler. This article was written by Rhodes Fairbridge, a professor at Columbia University. The article was written for the encyclopedia *Encarta*. I found the article difficult to read because I know very little about geomorphology. However, throughout the article many of the terms have links. These links were a great resource in learning more about the principles of geomorphology. Due to the prior knowledge needed for this article I would suggest that it be used for high school students and teachers. The article is separated into five different sections. These sections include Introduction, Historical Geomorphology, Process Geomorphology, Underlying Dynamics, and Weathering and Erosion. Each section offers a brief paragraph, along with photos and examples. I found the site to be easy to use and very informative. In addition this site also offers links to articles on similar topics.

Title: Fundamentals of Physical Geography Chapter 11: Introduction to Geomorphology
URL: http://www.geog.ouc.bc.ca/physgeog/contents/chapter11.html
Grade Level: Advanced placement high school or college
Best Search Engine: www.google.com
Key Search Word: geomorphology
Review: This site was created by Michael J. Pidwirny, Ph.D. at Okanagan University College. It was set up with the following main headings: Contents, Glossary, Study Guide, Links, Search, and Instructors. It also had subtopics from A to R. These topics included many areas, such as models of landforms development, weathering, introduction to soils, and soil classification. However, letters N, O, and P were of interest for this search. Once you clicked on "N. Introduction to Glaciations," you were able to read an introduction and the types of glaciers found. In "O. Glacial Processes," the highlights were growth of glaciers, glacier movement, and glacier mass balance. Finally, in "P.

Landforms of Glaciation," topics such as glacial erosion and glacial deposition was discussed. All of the information on this Web site was well organized and outlined. Each area of interest was full of very in depth explanations, as well as color photos, charts, graphs, and figures. The set-up was user friendly. In the text in each area, several words were highlighted in blue. When you clicked on these words, you got the definition of that word/phrase. I found that to be a helpful aspect. I also found that being able to view other links and search within the topic was helpful. The creator of this site also provided an e-mail address where he can answer questions. This site can provide the user with a great deal of knowledge on this subject.

Title: International Associations of Geomorphologists (AIG)
URL: http://www.geomorph.org/main.html
Grade Level: High school and college
Best Best Search Engine: www.mamma.com
Key Search Words: geomorphology association
Review: This AIG Web site serves as a comprehensive portal to support its members in their professional study of geomorphology as well as a means to introduce geomorphology to a broad audience. This searchable site is organized into six main sections: 1. About; 2. Members; 3. Working Groups; 4. Publications; 5. Meetings; and 6. Geomorphological Topics. Each of the six sections, accessible by the left-side navigation bar, provides links to relevant information and Web resources. The About and Members sections provide information regarding general membership and the organization. Working Groups highlights current committee work specific to arid regions, bedrock rivers, geoarchaeology, geomorphologic sites, large rivers, terriors viticoles, volcanoes, and previous working group topics. The Publications link offers archived issues of the *IAG Newsletter* dating back to March 1995 as well as access to the Geomorph listserv with archived postings available to nonsubscribers. Meetings provides information regarding upcoming international, regional and thematic meetings. For the student of geomorphology, the Geomorphological Topic section may prove the most useful. The Regional, Glossary, Image Gallery, and Links subsections provide a guide to classic landforms of the world; several interactive glossaries of geologic terms; a gallery of nondownloadable landform images; and a number of quality geomorphologic links organized by geomorphologic societies and groups, related organizations, and mailing and discussion groups on geomorphology. Overall, this site is authoritative in its content and timely in its coverage,

updated frequently by scholars in the field. On both the instructional and professional levels, the AIG Web site is a must visit.

Title: Geomorphology Journal
URL: http://www.sciencedirect.com/science?_ob=journal
Grade Level: College and above
Best Search Engine: www.metacrawler.com
Key Search Word: geomorphology
Review: The *Geomorphology Journal* Web site provides its most current research journal articles on geomorphology and Earth Science. The Web site maintains the professional integrity it demands of its published authors. Despite the high standards, the site is systematically organized, user friendly, and aesthetically appealing. The site lists journal volumes 11–14, four issues per volume, and it includes articles in press (52) from December 2001 through September 2002. Science Direct has divided the Web site into areas that include links to Home, Publications, My Alerts, My Profile, and Help. There is a register or login component requiring a name and password, and the Web site offers three types of usage: subscribed, unsubscribed, and complimentary. The most user-friendly feature is the list that conveniently pulls up volumes and issues by volume number, issue date, and year. I chose to review a summary of a 10-page article: "Ice Jam-Caused Fluvial Gullies and Scour Holes on Northern River Flood Plains" by Derald G. Smith and Cheryl M. Pearce of the Departments of Geography/Canada. The summary contained an abstract, which provided even a layman with a general understanding of what the article was about and what geomorphology is. It also explained the reason for such study and the potential application of the findings. The author's keywords guide was included, as well as an outline of the article, acknowledgements, references, DOI (Document Object Identifier) Scheme, and figures. This particular article made use of an index map, a bar graph, comparative drawings, aerial photos in black and white, and schematic models for its figure illustrations. Its tables consisted of values in meters and were cross-referenced to the text. The *Geomorphology Journal* Web site systematically presents research-based information with accompanying data on the study of changes in the Earth's landforms through naturally occurring events. It is updated issue by issue. As I am inadequately trained in the discipline of geomorphology I am not qualified to evaluate the actual research of the articles. I am, however, able to comment that this is a Web site worthy of the field and useful to someone doing research in the field. The language of the

articles is sophisticated and does not include "smoke screening" or basic elementary explanations. There is listed use of instruments of physical science noted within the text, which aids in research reliability. References are varied and include seminar material, surveys, theses, professional papers, technical reports, CD-ROM computer file data, research reports, journals, and books. The authors put themselves at risk for honesty by posting their phone and fax numbers and postal and e-mail addresses in order to receive reader feedback.

Glacial Geology

Title: The Geography Exchange Resource Centre
URL: http://www.zephryus.demon.co.uk/goegraphy/resources
Grade Level: Grade 6 and up
Best Search Engine: www.metacrawler.com
Key Search Words: glacial geology
Review: Good site on all of geology including plate tectonics, earthquakes, and volcanoes, but also one of the few sites I found that had any decent amount of information on glaciers. This site is specifically aimed at elementary and early high school students and provides a complete discussion of all kinds of glaciers and land formations created by glaciers. it contains definitions, descriptions, and an excellent array of supporting photography. I would highly recommend this site as an educational tool for not only glacial geology, but also all geology.

Title: The National Snow and Ice Data Center's All About Glaciers
URL: http://nsidc.org/links/glaciers.html
Grade Level: Grade 5 and up
Best Search Engine: www.dogpile.com
Key Search Words: glacial geology
Review: This subsite is part of the National Snow and Ice Data Center's Web site. While on the surface it doesn't have the glamor and glitz of other sites I accessed, there is a wealth of information presented in a straightforward manner. The site appears to be updated on a regular basis and the information provided is from a trustworthy source. The Web site is easy to navigate, but it is easy to get lost in all the links provided. The entry page has six bulleted links: glacier facts, questions and answers, a glacier glossary, a glacier gallery, glacier news, and links to glacier books, articles, and Web sites. Additional related links are underneath the bulleted list. The glacier facts page contains about 20 interesting facts about glaciers referencing size, shape,

length, and locations. This page has no link back to the previous page, and you must use your browser's back button. However, there is a menu at the top of the page with links to the other five links referenced on the entry page. The questions and answers page has a series of common questions along the left side of the page. You simply click on the question you wish answered and the answer appears. Again, there is no glitz, and the question is answered in a straightforward manner. The glacier glossary page has numerous glacier-related terms listed in alphabetical order along the left side of the page. Clicking on the term produces its definition in a frame to the right. One feature that is beneficial here is that if another glacier term is used in the definition, you may also click on that term to get its definition. The glacier gallery page has a list of glacial features and types of glaciers along the left side of the page. Again, clicking on one of the terms leads to an immediate picture. Also on this page is a list of links that also contain images of glaciers and maps relating to glaciers. It appears this page is updated on a regular basis—another plus for this Web site. The links include an educational materials link, which lists extensive resources for teachers. This would be a worthwhile Web site to include in a classroom study of glaciers.

Title: Connecticut Geology
URL: http://www.wesleyan.edu/ctgeology
Grade Level: Middle and high school
Best Search Engine: www.webcrawler.com
Key Search Words: glacial geology
Review: This is an excellent teacher resource with suggested lessons, extended links to related topics such as marine geology and the universe, and maps.

Title: The National Snow and Ice Data Center
URL: http://nsidc.org/glaciers
Grade Level: Elementary to college
Best Search Engine: www.yahooligans.com
Key Search Words: glacial geology
Review: This site is part of the National Snow and Ice Data Center. The creators of "All About Glaciers" promise the site has something of interest for everyone, from grade school children to geologists. The site has four sections: 1. Data and Science—this section provides links to information about glacial research, projects, and online organizations. 2. General Information—here you will find basic glacier facts,

FAQs, a glossary, pictures, and links to other glacier sites. This part of the site is easy to navigate, as the phrases/words reveal additional information and pictures when clicked. The glossary contains a lengthy list of words related to glaciers. The definitions are simple and easy to understand. 3. Glacier News—in this section you will find updates on the latest glacier occurrences. Happenings are listed beginning with the most recent in August 2003 back through June 1997. This section provides a brief overview of the article, followed by a link to the publications in which the story originally appeared. 4. The Glacier Story—by clicking here you are taken on a tour of a glacier's "life." There are nine pages to the Glacier Story, and each includes detailed information and great pictures. This site is easy to navigate. The pictures are clear and accompanied by great descriptions. I would definitely recommend this site to anyone studying glaciers.

Title: Glacial Geology
URL: http://members.aol.com/scipioiv/glacialgeology.html
Grade Level: Grades 7–9
Best Search Engine: www.37.com
Key Search Words: glacial geology
Review: This Web site is fundamentally sound and an excellent starting point for a beginner researching glacial geology. The article is approximately 4 pages long and contains a brief description of 22 terms associated with glacial geology, and in some cases a brief example of each. The two main types of glacial geology discussed are alpine and continental. Following the article there are nine pictures with an explanation as they directly relate to the reading, thus helping the reader further relate to the material they just read.

Title: Education World Web directory: Science: Earth Sciences: Geology: Geomorphology
URL: http://dirs.educationworld.net/cat/26962
Grade Level: Grade 9 and up
Best Search Engine: www.metacrawler.com
Key Search Words: geomorphology and glacial geology
Review: Education World serves as the main Web site and lists several Web sites in geomorphology. This site seemed user friendly and encourages the user to explore several resources about geomorphology. This site features two dozen subcategories in the field of Earth Science. The Web sites seemed to be accessible through this main Web site. This Web site could be beneficial to a glacial geomorphology re-

searcher. The main page offers a collection of photos, short articles, and bulletins to explore. I chose to explore a subsite called The Alfred Wegener Institute, Germany's leading institute for polar and marine research. This Web site lets users explore the recent research written in German and English. The Web site seems to have interesting facts regarding the climate system, types of ecosystems, and ecosystems. The site features the links Research, Resources, Click and Learn, About Us, and Ships and Stations on every page. For example, Click and Learn features frequently asked questions designed for children. The list is updated often through the institute. The children are provided with answers regarding ways to contact the board with a nominal fee. The differences between the Arctic and Antarctic are explained in this subsection. The children are also informed of avenues they could explore if they were to become a marine biologist or polar scientist. You can simply click back to About Us to return to the main page. I found this to be helpful with any Web site: clicking back to the home base seems to take me to the main page in English. This Web site can provide information to a researcher, educator, or student interested with glacial geology.

Title: Glacial Geology
URL: www.uoguelph.ca/~sadura/glref/gl0.htm
Grade Level: AP high school and college
Best Search Engine: www.metacrawler.com
Key Search Words: Glacial geology index
Review: This site was developed to clarify the terms and concepts related to glacial geology. Steven Sadura, a professor at the University of Guelph, created the site. The site is divided into 10 categories. The categories are: 1. Introduction, 2. Glaciations, 3. Settings, 4. Properties of Ice, 5. Movement of Glaciers, 6. Classification of Glaciers, 7. Erosion, 8. Transport, 9. Deposits, and 10. Effects of Glaciations. The learner will click on a category and view a short slide show to find information about the selected topic. Almost all of the slides include pictures, maps, charts, and graphs to help the learner understand the terms and definitions presented. The slides are presented in a clear and easy to read manner. The best part about the slides was that they were short and to the point. Most slides did not use more than 20 words to convey information about glaciers. However, I do feel the user may need prior knowledge about glaciers before reading this site. Some information was hard for me to comprehend due to my lack of knowledge about glacial geology. The most extensive categories in the

Web site are glaciations, settings, and movements of glaciers. These categories offer numerous slides with more terms and pictures than the other slides on the site. The site offers a link to glacial geology resources on the Web. This link gives seven sites that may help the user to extend his or her knowledge on glacial geology. The site also links back to the creator so the user can e-mail any questions or comments about glacial geology. I feel this user-friendly site is effective in describing the terms and concepts associated with glacial geology.

Title: Glacial Geology Geomorphology
URL: http://boris.qub.ac.uk/ggg
Grade Level: Grade 10 and up
Best Search Engine: www.metacrawler.com
Key Search Words: glacial geology
Review: The link to the page off of the Metacrawler search results was GGG Homepage. The link, to me is a little misleading. It holds some information, but is still in the process of being built. The page is appropriate for sophomores in high school and up only. Started by a research group from The Queens University at Belfast, the language and setup is very professional. I would suggest it only for the links page. There is a multitude of useful links to see other research done about the topic. Also, there is a pictures page showing examples of the topic studied. These are clear and well organized. The resources link also leads you to newsgroups and paper journals you can track down. The bulk of the site is about electronic resources, though. One of the biggest letdowns was the diary link. The entries stopped three years before my visit. This could mean that the professor does not work at the school any more, but one would think that it would be taken over. Overall, I would recommend the site purely as a source of links to other information.

Title: A Hypertext in Glacial Geography
URL: http://www.homepage.montana.edu/~geol445/hyperglac
Grade Level: College
Best Search Engine: www.metacrawler.com
Key Search Words: glacial geology
Review: This is a great page for those who are looking for a comprehensive introduction to glacial geology. Created by a geology class at Montana State University, this Web site is constructed so that navigation is easy. The overview page has click to enlarge thumbnail pictures giving examples of what is being discussed in the adjacent text.

There are many links available to let you navigate the site without having to backtrack. There are, however, a few external links that appear to be dead. Moreover, the internal link that should give us references does not work. Even with these flaws the site is a great resource for those who wish to learn about glaciers and the geology associated with them.

Title: Origins of Niagara: A Geological History
URL: http://www.iaw.com/~falls/origins.html
Grade Level: High school
Best Search Engine: www.metacrawler.com
Key Search Words: glacial geology
Review: This Web site was interesting to read and review. Although it is mainly all about Niagara Falls, there is a bit of interesting information about glaciers. It begins with the Table of Contents, which includes such topics as Geography of Niagara, Niagara Escarpment (which includes a timeline), Wisconsin Glaciers in Niagara, Birth of Niagara Falls, and others. The Wisconsin Glaciers topic explained how the glacier retreated to eventually form the Great Lakes. This was basically the only reason that this site came up under glacial geology. There are probably other sites more worthy of your time. There are few links of interest to the glacier topic, but the author did include his e-mail address for any questions about the Web site and Niagara Falls in general. In conclusion, there are probably other sites more worthy of your time (unless you are especially interested in Niagara Falls).

Title: Glacial Geology
URL: http://www.homepage.montana.edu~geol445/hyperglac/index.htm
Best Search Engine: www.earthlink.net
Key Search Words: glacial geology
Review: The hypertext site was developed to give an appreciation of glaciers and how they work. The title Glaciology and Glacial Geology is the first thing you observe when entering the site. The students at Montana State University with the help of their professor created the site. The home page gives a quick overview of the organization and how to navigate the links. The Table of Contents is divided into four sections: Overview, Introductory, Advanced and Ancillary. Each section is in bold print, offering a different color from the background and the paragraph on the organization. The Overview discusses glaciers 101. As you navigate through this page you find small pictures on the left and arrows pointing up to glacial erosion. Each word is in

bold print and gives a paragraph with information for example: mass budget, glacier erosion, glaciers, glacial erosion-fiords, and trim lines to name a few. The Introductory and Advanced Table of Contents are identical in their 12 subtopics. The Introductory links give you subtopics on Depositional Landscape, Glaciers with Time, Systems: Budget and Flow, and nine other topics. Each of the links provides colorful pictures with information and sites to link to more information on the topic. The Advanced Table of Contents has drawings and few color pictures. Each subtopic has a more detailed description of the information given as well as other links. The text used with the background on some of the subtopics is smaller than the introductory section and harder to read. The Ancillary contains a glossary, references, and permissions. The Glossary is in alphabetical order, starting with Ablation and ending with Valley Train. The words are not highlighted. The References link is not available and the Permission link is not accessible. The site is quite extensive and gives you information whether you are a beginner or an advanced student. It provides many links and is easy to navigate.

Hurricanes and Cyclones

Title: CIMSS Tropical Cyclones
URL: http://cimss.ssec.wisc.edu/tropic/tropic.html
Grade Level: College or specialist in the field
Best Search Engine: www.yahoo.com
Key Search Words: hurricanes and cyclones
Review: This page is from the University of Wisconsin–Madison. CIMSS stands for Cooperative Institute for Meteorological Satellite Studies. Everything on this front page seems to lead you to another link. One takes you to the university; the other takes you to the CIMSS home page. If you click on Jason Dunion's name, it takes you to his page, which appears to be a bibliography of articles he has written. Most of the other names are inoperative, so their individual pages may be under construction. The CIMSS page seems to be the most interesting. It has links to operational satellites, research satellites, ground-based sensors, airborne sensors, numerical weather prediction, the weather, field experiments, interdisciplinary projects, and outreach and education. Within each category, there are several links you can choose. I will choose one from interdisciplinary (hey—there is a link to a geomorphology site). It is called "coastal Louisiana geomorphology." This talks about the erosion on the Louisiana coast. There are

several links here. (I think, in my amateur opinion, that this may be a good site to use due to the fact it ties in geomorphology; http://cimss. ssec.wisc.edu/clageo.) This Web page has a "what's new" section, a publications section, an image gallery, data products, related links, and some other links. I recommend further investigation of this site as well as the Louisiana site. They seem to tie together nicely.

Title: American Meteorological Society
URL: http://www.ametsoc.org/ams
Grade Level: High school and college, teachers
Best Search Engine: www.metacrawler.com
Key Search Words: hurricanes and cyclones
Review: This Web site is a journal devoted to weather and weather-related issues. At first I found it difficult to navigate the site. As I became more comfortable I discovered many valuable resources. The site lists current workshops and seminars being held around the country on weather. It describes current weather around the globe. I found many interesting articles and facts about hurricanes and cyclones. In addition the site offers an area where you can search though previous issues to find a particular topic. This site also offers a glossary of terms you can access for teaching purposes. I found that this Web site would be too advanced for elementary school children. The site does, however, offer a wide variety of resources for older students and teachers.

Title: Atlantic Oceanographic and Meteorological Laboratory
URL: http://www.aoml.noss.gov/hrd/weather-sub/faq.html
Grade Level: High school and adult
Best Search Engine: www.alltheweb.com
Key Search Word: cyclone
Review: This site has subsites with site maps, staff, data center, contact information, and a research division. Once inside the site you can go from basic introduction and definition to cyclone names, myths, winds, records, forecasting, climatology, and observation. If you want to get deeper into the information there is real-time information and historical information. They also give information to help you prepare for a hurricane. In the end there is a reference guide.

Title: Hurricanes: Online Meteorology Guide
URL: http://ww2010.atmos.uiuc.edu/(Gh)/guides/mtr/hurr/home.rxml
Grade Level: Grade 5 through high school
Best Search Engine: www.yahoo.com

Key Search Words: education and hurricanes

Review: The Hurricanes: Online Meteorology Guide Web site was developed by the Department of Atmospheric Sciences at the University of Illinois and is part of the larger "WW2010" operation, which is an educational site on meteorology. The hurricane page provides 12 sections to explore on specific topics. These sections include: Definition and Growth; Stages of Development; Structure of a Hurricane; Explore a 3-D Hurricane; Movement; Satellites and Hurricane Hunters; Preparations; Damage and Destruction; Hurricane Tracks; How They Are Named; Global Activity; and El Niño. Each of these sections includes specific information appropriate for middle school and high school use. Each section also includes colorful pictures and graphics (some of which show movement and progression). Information is also provided in chart form. All graphics have been developed in a way that attracts attention and makes them interesting to study. With additional multimedia technology one can also fly through a 3-D hurricane and interact with hurricanes. Navigation was problem free, although upgrades are needed for two interactive parts of the site. Since this is a basic information site and is not overly cumbersome with information, it is a great resource for students working on reports, for teachers to let students explore in the classroom, and is an excellent resource for teachers to access graphics.

Title: Weather Watch: Hurricanes! Discover the Power and Fury of Tropical Storms

URL: http://teacher.scholastic.com/activities/wwatch/hurricanes/index.htm

Grade Level: Grades K–8

Best Search Engine: www.yahooligans.com

Key Search Words: hurricanes and cyclones

Review: As part of Scholastic.com's online activities center designed for kindergarten through eighth grade teachers and students, Weather Watch is an ongoing series of weather-centered online science projects. Hurricanes! Discover the Power and Fury of Tropical Storms is the first of the series of extreme weather systems highlighted on this site. Through a variety of interactive learning experiences and online resources, students can use this site to learn about the features of a live hurricane and how to track its path. This site is organized into three main sections: All About Hurricanes, Ask Our Weather Expert, and Hurricane News. The All About Hurricanes section answers the question "What is a hurricane?" and provides instructions to use a hurricane tracking map and to plot a hurricane's position. Students can access a downloadable tracking map as well as links to national and international weather centers that provide current and historical data on tropical storms, animated tracking maps, and live satellite images. This section also provides links to hurricane trivia and two structured Internet field trips for grades pre-K through 3 and grades 4 through 8. The Ask Our Weather Expert section provides access to National Weather Service meteorologists for online question and answer sessions as well as archived interviews. The Hurricane News area provides a news story that predicts the intensity of the current hurricane season. For additional research on hurricanes and cyclones, this site also provides a link to the Extreme Weather Research Starter Center, where students can use the online glossary to look up weather related terms like cyclone and hurricanes and gain access to a variety of online weather-related reference tools and articles. In addition to the instructional resources for students, this site also provides a teacher's guide for grades K–8. The guide, whose links appear on the left side, includes a project description, an assessment and rubric section, learning objectives, project components, lesson planning suggestions, national standards correlations, cross-curricular extensions, and additional related resources, including books and software. Overall, this easy-to-use site provides clear content and well-organized links to hurricane and cyclone resources that will support an elementary to middle school student's study of hurricanes and cyclones.

Title: Hurricane Research Division: Frequently Asked Questions
URL: http://www.aoml.noaa.gov/hrd/tcfaq/tcfaqHED.html
Grade Level: Advanced middle school, high school, college
Best Search Engine: www.google.com
Key Search Words: hurricanes and cyclones
Review: This advanced level Web site offers a wealth of information for
    both high school students (perhaps if you are taking an oceanography
    course) or college students. This information is available in English,
    Spanish, and French. It is managed by the Atlantic Oceanographic
    and Meteorological Laboratory (AOML) division of the National
    Oceanic and Atmospheric Administration (NOAA). This laboratory
    is in Miami, Florida, and is part of the U.S. Department of Com-
    merce. This is an excellent site for researching topics such as hurri-
    canes, El Niño, and current projects being overseen by NOAA. The
    research division includes information on Hurricane Research, the
    Ocean, and Chemistry, as well as Physical Oceanography. The Web

site consists of various definitions, answers for some specific questions, and information about the various tropical cyclone basins. Numerous publications are available on the subtopics of Ocean and Climate, Coastal and Regional, and Hurricanes. These publications include journals, books, theses and dissertations, and technical and data reports. It also includes numerous related sites for further research. It also provides sites where you can access real-time information about tropical cyclones, find information and resources for historical storms, and get lists of books to read and references for tropical cyclones. The main purpose of the Web site is to provide quick answers for frequently asked questions as well as to provide information about various weather conditions. This is an extensive and thorough Web site where one could easily spend hours learning about weather conditions. The site is loaded with pictures, charts, graphs, tables, and detailed written documentation. I chose to look into "Real-time Monitoring of Tropical Atlantic Hurricane Potential for 2002." The site is updated from June through November. Statistics are provided, which predict the probability of storms along the Atlantic. Charts depicting water temperatures and wind conditions are included. My favorite site was "Hurricane Awareness." The National Oceanic and Atmospheric Administration, American Red Cross and the Federal Emergency Management Agency were participating organizations. The site answered questions and provided in-depth information related to what a hurricane actually is, threats, who is at risk, what to listen for, safety issues, a family hurricane plan, Florida statewide awareness events, and hurricane intensity scales. This entire Web site is well thought out and descriptive. It is user friendly, and the photographs and charts add to the comprehensive text presentations. This would make an excellent tool for someone studying weather who may want to gather information for a research paper. Additional references are recommended on the Web site so people who are interested can expand on the topics of interest. It is a superior government site and well suited for the advanced learner as there is a wealth of current scientific information available.

Hydrology

Title: The Hydrologic System
URL: http://geography.about.com/gi/dynamic/offsite.htm
Grade Level: Grade 9 and up
Best Search Engine: www.metacrawler.com

Key Search Word: hydrology

Review: This site is produced with information from Environment Canada. This comprehensive site has numerous links that are easy to navigate through. The site is divided into nine main topics: 1. Introduction, 2. The Hydrologic Cycle, 3. Quick Facts, 4. Water here, there, and everywhere, 5. Rivers, 6. Lakes, 7. Groundwater, 8. Glaciers, and 9. Snowfall. Within each main topic there are many links for subtopics. The best part of the site is the link for the hydrologic cycle. Each part of the hydrologic cycle is explained in an easy to read manner. This link also has a great graphic that illustrates the hydrologic cycle. Another neat feature is the link for quick facts. This page gives the users many interesting facts about water. Any high school student looking for information ranging from the properties of water to the hydrologic system could easily find it on this great Web site.

Title: Global Hydrology Resource Center

URL: http://ghrc.msfc.nasa.gov

Best Search Engine: www.earthlink.com

Key Search Word: hydrology

Review: The site is eyecatching and full of information on hydrology. The background has a drop of water splashing up, which seems to catch your eye. On the left side of the page are 13 items to choose from: GHRC home page, project information, glossary of terms, general information, related sites, feedback, hydrology news, e-mail, feedback, and a few more. In the middle of the page are four paragraphs that give information on four areas. The first paragraph discusses the Global Hydrology and Climate Center in Alabama. Once you click into the site it gives information on conditions in Alabama, data, education, research, and climate impact. The paragraph on lighting discusses the cause and effect of lightning and analyzes a variety of atmosphere measurements related to thunderstorms. That link gives information on primer, dataset information, research, observations, validation, and books. The paragraph on the Convection and Moisture experiment gives a link to information on research investigations sponsored by the Earth Science Enterprise of NASA. The link provides images and lots of information. The third paragraph gives a link to the Passive Microwave Earth Science Information partner. This link allows the researcher to interactively customize and receive hydrologic data from the new space-based passive microwave instruments. The site gives a lot of information and links to find the data. Beginners or advanced computer people can easily navigate the site. The feedback page allows you to give comments on the site and ask questions. The site may look overwhelming, but is put together nicely.

Title: Napa County Resource Conservation District
URL: http://www.naparcd.org/hydrology.htm
Grade Level: Grades 11 through college; elementary school teachers
Best Search Engine: www.webcrawler.com
Key Search Word: hydrology
Review: If you're looking for a Web site that you can use to teach across the curriculum, that is, civics, geography, science, and history, this is an excellent site. It highlights the thorny issue of water flow through an area that has a history built around water rights. Most of its links would be beneficial to the secondary ed teachers, but the education link has photos of children experiencing testing and conservation firsthand. Ergo, the upper elementary teachers might find that link useful. Most of the vocabulary was comprehensible and not too technical, to this writer. This is one of the main reasons for recommending this site. The other is as I mentioned before: it serves as a great way to teach across the curriculum. I can envision a nice little essay paper from high school students on the history of water rights in sections of California and the southwest.

Title: MetEd Meteorology Education & Training: Hydrology
URL: http://meted.ucar.edu/resource/metlinks/hydrotut.htm
Grade Level: Grade 6 and up
Best Search Engine: www.metacrawler.com
Key Search Words: hydrology education
Review: This site has two good tutorials on making a river forecast and making a headwater forecast. It involves a lot of science and math, but it is a good interdisciplinary site for those two subjects. The site provides many definitions for new and unfamiliar vocabulary, and good explanations of why such forecasts are needed. The processes involved in making these forecasts are described very thoroughly, and they make great use of the scientific method (of which I am a big fan). There is also a couple of great examples of flowcharts outlining the forecasting processes. Overall good site. May need to be thoroughly previewed by instructor as there is some difficult material that may need to be simplified.

Title: Center for Ecology and Hydrology, Natural Environment Research Council
URL: http://www.ceh-nerc.ac.uk
Grade Level: College/post graduate
Best Search Engine: www.dogpile.com
Key Search Word: hydrology
Review: This Web site claims to be the "leading body in the UK for research, survey, and monitoring in terrestrial and freshwater environments." The site has five links: News, About Us, Science, Products, and Data. In News, you will find recently updated research information and upcoming events. About Us provides links to the various facilities of the organization throughout the United Kingdom. When you go to Science, you will find information about the types of research projects of the organization; soil-vegetation, land-use urban environments, freshwater quality, biodiversity, biocontrol, pollution, extremes, and global change. At the end of the page, a contact name/e-mail address is given. Products gives you information regarding the various products and publications of the organization. There are three links on this page: Software Products, Publications Office, and Library Services. The Data page contains a wealth of information. There are links to live databases, and a landcover map of Great Britain, which contains data from satellite data collected by Landsat 5's Thematic Mapper. The Links page is filled with links to related sites. The page is easily navigated and includes the following links: soil-vegetation, land use urban environments, freshwater quality, biodiversity, biocontrol, pollution, extremes, and global change. There are not too many graphics on this site, but it does include a lot of information and related links. I think this site will be of great use to researchers in the field.

Title: World Wide Web Pages for Dam Design
URL: http://www.dur.ac.uk/~des0www4/cal/dams/fron/contents.htm
Grade Level: Grade 6 and up
Best Search Engine: www.yahooligans.com
Key Search Word: hydrology
Review: This is one of the most straightforward, best-organized Web sites I have ever seen. Educators and researchers from the University of Durham started this Web site in 1997. Each section is separated into other sections under a well-organized outline. There is a glossary that defines every term used on the site in a clear and concise manner.

The help section explains the entire site, using examples and visuals to aid the visitor. I would recommend this site to science teachers to use in their classroom. This is a great resource of information and visual aids.

Title: The Hydrosphere and Hydrologic Cycle
URL: www.uwsp.edu/geo/faculty/ritter/geog101/modules/hydrosphere/
    hydrosphere_title_page.html
Grade Level: Middle school through college
Best Search Engine: www.yahoo.com
Key Search Word: hydrology
Review: This Web site is actually for a Geography 101 class at the University of Wisconsin–Stevens Point. The title page and the table of contents are excellent resources. The information is extremely easy to read. The contents consist of Hydrosphere and Distribution of Water, Hydrologic Cycle, Subsurface Water, Surface Water and the Water Balance. There are excellent graphics of the water cycle. Under Hydrologic Cycle, evaporation, precipitation, infiltration, and interception are defined and explained. There are graphs showing the difference between available water and soil texture. There is also a section where the students taking this course can assess themselves. Under Water Balance, there are important definitions. This looks like a great site to obtain the basic needed research.

Title: American Institute of Hydrology
URL: http://www.aihydro.org
Grade Level: Grade 8 or above
Best Search Engine: www.dogpile.com
Key Search Word: hydrology
Review: This site is a site given totally to the subject of hydrology. It is a site with 12 links. Its home page starts out by informing the reader of upcoming conferences on hydrology. The first link, Hot Topics, gives the reader the latest updates on hydrology. The links continue on with information about the Hydrology Institute and information how one can become a member. This site next deals with hydrology examinations performed or to be performed. The links continue with a calendar of events on hydrology. It further discusses how one can subscribe to the institute's publications. Another link goes to a state section. The next link includes a section for students. Other links are given for additional research. Finally there is a link for corporate

members and corporations that deal with hydrology. I think this site is perfect for one who wishes to research Hydrology. The site is designed to assist hydrology workers and researchers.

Title: Hydrology
URL: http://www.em.doc.gov/soda/indx.html
Grade Level: Elementary and middle school science teachers
Best Search Engine: www.metacrawler.com
Key Search Word: hydrology
Review: This site was prepared by the U.S. Department of Energy Office of Environmental Management. It opens with a heading, Soda Bottle Hydrology, and an invitation to science teachers to try out activities designed to help children understand the concepts of hydrology. The page continues with an explanation of hydrology, a teacher's introduction explaining the purpose of the activities listed, a student's introduction, a list of materials needed for the activities, and 12 activities. Clicking on the title of an activity takes you to an experiment meant to explain that concept. To get to another activity you can click on a button at the bottom of the page or you can return to the home page. At the end of the home page there are links to other sites. These links include EM home, DOE home, Search, Website Outline, Feedback, Accessibility, Privacy, and Security. Clicking on DOE Home takes you to the home page for the Department of Energy. It contains links to many additional sites that have nothing to do with hydrology. Clicking on Search leads you to a link to search either all of the Department of Energy or just the Office of Energy Management. The Website Outline is 15 pages of topics related to the Department of Energy Management, some of which are Featured Items, Press Releases, Learning about the Department, EM Programs, and on and on and on. While this site did contain some activities on hydrology that science teachers might find useful, it didn't appeal to me. Once past the activities, the links seemed to take you all over the map. The site is easy to navigate; however, it would take many hours to investigate each link and few if any links dealt with hydrology. I would think the information on hydrology in this site could be found in a good bookstore.

Hydrosphere

Title: Hydrosphere Gateway Page
URL: http://www.hydrosphere.com
Grade Level: Field experts and specialists

Best Search Engine: www.yahoo.com
Key Search Word: hydrosphere
Review: The opening page offers two ways: Resource Consultants or Data
    Products, Inc. Going into the resource consultants' section, there are
    a few links to the left: Company, Service, News, Projects, Opportuni-
    ties, Contact, Home, and Us. There is also an icon for HRC GSA En-
    vironmental Services. Also this home page lists the types of services
    available, such as watershed studies, National Environmental Protec-
    tion Agency compliance documentation, water conservation studies,
    rainfall runoff, wildfire management and mitigation planning, data-
    base design and development, and data management. The project
    page lists current projects and has a few links you can go to for addi-
    tional information. Software can be downloaded under Technical
    Support. There are demonstration versions that can be tried, and a
    link to other cyber environmental links. The site also has a section
    for upcoming conferences. This site does not give access to other re-
    lated links, aside from HRC GSA Environmental Services. That link
    supplies only contact information. This site is specific to its company
    and the company's purpose.

Title: Environmental Chemistry
URL: http://www.ul.ie./~ces/hydrosphere.htm
Grade Level: High school to college freshman
Best Search Engine: www.google.com
Key Search Words: hydrosphere, hydrology
Review: This hydrosphere Web site comes up titled Environmental
    Chemistry. At the top are windows labeled The Hydrosphere, Atmo-
    sphere, Lithosphere, Glossary of Terms, Search, and FAQ. Under-
    neath are the following topics:

- Intro to the hydrosphere: looks at properties of water, the water cycle,
  and lakes.
- Acid-Base equilibria: the importance of acid-base equilibria in water.
- Complexation and Chelation: the formation of metal complexes in
  water and their stability and the chelate effect.
- Redox-electrode potential diagrams: electrode potentials and drawing
  distribution diagrams.
- Solubility/precipitation equilibria: common ion effect, precipitation, and
  more diagrams
- Heavy metals and waste treatment: waste treatment, water quality, pollution and hardness in water.

Nowhere does the Web site say who developed it. There is no university logo or organization name. References listed at the bottom of each topic/essay link to a picture of the teacher or author of the essay, list credits, and state the goal of the essay. The goal of the intro to the hydrosphere essay is to train students to understand and solve real-world environmental problems through basic hands-on science activities. There is an overview section that links to staff, writers, collaborators, and an advisory panel. Each topic/essay consists of a table of contents, and introduction, case study, essays, and an experiment. The explanations are technical, and fortunately they use glossary links on the technical terms. Great use is made of pictorials to detail the distribution of water. The acid-base equilibrium essay includes chemistry formulas and formula applications. There is a link to a "general chemistry" site with a glossary, a general chemistry review, and the periodic table. There is a Quick Test of your knowledge on acids and bases, and a tutorial on drawing distribution diagrams. This could be helpful in illustrating one's own project on this topic. The topic/essay for complexation and chelation has similar features. The Redox equilibria and Pe-Ph diagrams topic/essay is under construction. There is a one-paragraph introduction to the topic. There is a link to the "best available" site on this topic, but when you click on it, it says that the material is outdated and has been removed as it has many broken links. If you desire this information, you are urged to contact Dr. Plambeck, but it adds that Dr. Plambeck has retired and is not obligated to respond. Overall I would feel comfortable using this Web site for information about the hydrosphere and topics appertaining, however, I would not feel comfortable quoting any of the information as reference unless it specifically cites who the author(s) is and where the information generates from. Only two of the topics offered this information.

Title: Hydrosphere
URL: http://earth.rice.edu/mtpe/hydro/hydrosphere.html
Grade Level: Grade 5 through high school
Best Search Engine: www.mamma.com
Key Search Word: hydrosphere
Review: This well-organized Web site is a component of the Public Connection, a collaboration between Rice University and the Houston Museum of Natural Science. The site answers the five basic questions: 1. "What is the Hydrosphere?," 2. "Why do we care about the Hydrosphere?," 3. "How do we study the Hydrosphere?," 4. "Who are

the people interested in the Hydrosphere?," and 5. "Where can I get more information about the Hydrosphere?." The information presented is extremely straightforward and easy to navigate. The key words what, why, how, who, and where are links at the top of the site. When the learner clicks on the key word, the corresponding question is posed at the top of the page and supporting content and Web links are provided. The additional links, What's Hot? Sphere Topics, and Latest Images, also appear at the top of the page. Each link provides additional resources, which can be accessed via the links on the left-side navigation bar. For example, the Latest Images link provides links to images and maps addressing the following questions: How warm are the seas today? Is there a flood watch today? How fast are the winds? How much water is in the clouds? Is it raining in the tropics today? Are the seas high or low today? How high are the sea waves? In addition to the comprehensive information on the hydrosphere, links to resources on the atmosphere, geosphere, cryosphere, and biosphere are also provided. This site is a gem for any learner looking for well-presented information on the hydrosphere.

Title: The Open Door Web Site
URL: http://www.saburchill.com/intro.html
Grade Level: Grades 4–8
Best Search Engine: www.google.com
Key Search Word: hydrosphere
Review: This Web site is a wonderful asset for teachers, parents, and students. It is a Web site designed to search and provide information on a wide variety of topics. The site has a search button, which allows you to search through all of its information for a particular subject. The site is divided into five main categories. These categories are biology, chemistry, physics, technology, and history. I was easily able to find information on the hydrosphere. The Web site provided a description of the hydrosphere along with pictures. I found the material easy to understand and well explained. I think this site is an ideal for students who are searching for information for projects. The site encourages both students and teachers to use the information for projects and lessons. In addition the site provides an Internet Guide, which will help you find additional information on your topic. The site is well maintained and user friendly. I look forward to using this site in my classroom. It offers a wealth of information that will help make teaching certain subjects more enjoyable and understanding.

Title: United States Geological Survey—Science for a Changing World
URL: http://ga.water.usgs.gov/edu/mearth/html
Grade Level: Middle and high school
Best Search Engine: www.google.com
Key Search Words: hydrosphere and education
Review: This is a fun site full of great information for students, teachers, and
    your own children who have curious minds. Once you get to the
    "USGS—Science for a changing world" site you have a choice of edu-
    cation resources. You can go to USGS Learning Web, which provides
    links for students (reports and projects), teachers (lesson plans, models)
    and explorers (research tools and special topics). There is also a Search
    Wizard. For those looking for information on the hydrosphere you will
    click on Water Science for Schools, which provides information about
    everything you ever wanted to know about water using pictures, data,
    maps, and interactive sites. When you click on Earth's Waters you will
    see a menu featuring the following topics: Where is the Earth's Water?;
    How Much is There?; The Water Cycle; Water on Earth's Surface;
    Water in the Ground; Rain; and Glaciers & Icecaps. Each of these top-
    ics provides a tremendous amount of information. Each page includes
    links that help define terms the student may not be familiar with, in ad-
    dition to links to pictures that assist in comprehension of the subject
    matter. This site features colorful graphics, which are cute and keep the
    attention of the user. Some of the topics include an interactive feature
    that allows the user to take a true/false test and guess at some interesting
    facts. You can also vote for your favorite body of water. This site pro-
    vides links to many other sites. Users exploring this site can then search
    for additional information at a related site. At the bottom of each page
    is a glossary that the user can access if a term is not familiar. Even
    though this site is cute and interactive it is not suitable for younger stu-
    dents unless an adult is working with them due to the wealth of infor-
    mation provided. Younger students may get bogged down. This site is
    easy to use and problem free. There is an e-mail address to contact a
    person at the U.S. Geological Survey with comments.

Title: Hydrosphere
URL: www.miamisci.org/ecolinks/hydrosphere.html
Grade Level: All grade levels
Best Search Engine: www.ask.com
Key Search Words: what is the hydrosphere
Review: This is for students in any level of education. I believe it is put
    together by high school students, but to be used by all. It gives a de-

scription of what the hydrosphere is and how it works. There are also links to display student projects on coral reefs. There is a nice description of ocean currents, what causes them, and how and where they move. This link also includes a map depicting the currents around the globe. There are continued Hydro Links for Florida's Ecosystems, NOAA, and Water Purification. Most importantly this site explains how human beings can affect the world's water supply.

Title: Hydrosphere
URL: http://ess.geology.ufl.edu
Grade Level: Grades 6–8
Best Search Engine: www.google.com
Key Search Word: hydrosphere
Review: This educational Web site offers an excellent resource for research at the middle school level. It focuses on the Earth's hydrosphere with a brief description as well as a visual depiction. It also offers links for the following related topics: Atmosphere, Biosphere, Geosphere, and Anthrosphere (the last one being new to me, which I found astonishing as I used to teach anthropology and had never come across that term). All of these topics are depicted visually as well as described in text. It is most definitely too sophomoric for high school students but would serve as a decent source for reference for students grades 6 through 8.

Title: Hydrology Investigation
URL: http://www.globe.gov/sdabin/wt/ghp/tg+L(en)+P(hydrology/
    Contents)
Grade Level: Grades 7 and up
Best Search Engine: www.dogpile.com
Key Search Word: hydrosphere
Review: This subsite is part of the Globe Program, which provides hands-on science activities for primary grades through high school. This program allows students to interact with other students and scientists around the world to learn more about the planet Earth. The Hydrology Investigation is a chapter of a "web book" and is basically a resource for teachers looking to incorporate hands-on activities into the science curriculum. It could serve as a complete unit on hydrology, or specific activities could be incorporated into any science program. Not only are the activities laid out in easy to follow directions complete with diagrams where applicable, but also learner outcomes

as well as assessment options are included. There is even a suggested sequence of activities for teachers to follow. The Web page is divided into five sections: Welcome, Introduction, Protocols, Learning Activities, and an Appendix. Each of these sections contains links applicable to that particular section. For example, the Welcome section has a letter from two scientists involved in hydrology. The introduction contains links for student learning goals and assessment. The Protocols section contains links to a page that links to the activities for the students. Each Protocol link on that page links to specific, easy-to-follow directions for each activity. This is also true of the Learning Activities page. These pages include diagrams where applicable to aid in understanding the scientific procedure spelled out. There is also the opportunity for students to get data from specific locations submitted by students in other schools.

Title: Rader's Geography for kids
URL: http://www.geography4kids.com
Grade Level: Grade 3 and up
Best Search Engine: www.metacrawler.com
Key Search Words: hydrosphere and water
Review: This physical geography Web site is designed for children in grade three and up to explore. This site seems to be a beginner Web site to gain the basic knowledge of Earth Science and contains just about everything one would need to know about the hydrosphere. The site was created by Andrew Rader Studios and is featured as a popular Web site to explore in the education world by many professionals. The site's main page features many icons to open and explore. The icons are: energy, sky, land, earth, water, climate, free stuff, activities, expeditions, tour map/map, and examples. I was impressed with this site because it offered activities and free stuff for children. The site offers search functions, activities, downloads, and a search function on the left side of every page. There is a general descriptive page with bright, clear graphics. The general groups are water cycles, fresh water, ground water, seawater, and biomes. From there it breaks downs the topics to dense fog, freshwater, creek, high altitudes, lake, ocean waves, Pacific tide pools, storm clouds, stream mouth, and water tank. Through out the Web site there are definitions with key-term review. The most helpful way to navigate this Web site is to take a tour and see the site map. The site map offers users to become familiar with physical geography and to explore the different fields of physical geography. I showed this Web site to my niece when she was

over. She enjoyed this site and she is in third grade. She liked the activities, free downloads, and colorful pictures. She and I spent an amount of time exploring all the pictures and discovering so much information. She enjoyed the pictures and found the quizzes to be difficult. She was able to find information about the Earth and its hydrosphere. The Web site seems easy to navigate and easy to read. Bold words within the text address important points. There are illustrations on every passage for checking understanding. This Web site offers a lot of information and is updated often. I found this Web site to be the most interesting since it has pictures to clarify ideas.

Lithosphere

Title: Lithosphere
URL: www.duedall.fit.edu/sciweb/lithosph.htm
Grade Level: K–12
Best Search Engine: www.metacrawler.com
Key Search Word: lithosphere
Review: This lithosphere Web site was an interesting one, at least from my point of view. It has links to other sites containing information on other spheres, such as hydrosphere. Each link also had an evaluation, which allows you to rate the link. There were also names, addresses, and numbers to call for questions/problems.

Title: Lithosphere Dynamics and Continental Deformation
URL: http//earth.agu.org/revgeophys/bird01/bird01.html
Grade Level: High school
Best Search Engine: www.google.com
Key Search Word: lithosphere
Review: As you enter this Web site you will notice that it is very basic. There are no pictures and no colors. The site is from the Department of Earth and Space Sciences at the University of California, Los Angeles. In the top left corner there are four boxes: next, up, previous, and contents. Contents you can read about include major faults, pore water at high pressure, extension in the basin and range province, block rotation along the Pacific margin, Alaska and California neotectonics, huge displacement of the mantle, lithosphere, and references. If you go into the content boxes you will come up with headings like properties of solid earth, dynamics of solid earth and other planets, space sciences, atmospheric sciences, hydrology, and ocean sciences. If you click into these items they give you more information on these subjects and other

links. As you navigate through the 10 items you will get brief information on each of the options. In each item description there are reference links that lead to pages of references. The information is short and does not give you easily understandable material. This site does not allow you to link to other sites.

Title: Clues to Plate Movements
URL: http://www.pearsoncustom.com/link/lithosphere.html
Grade Level: Grades 7–12
Best Search Engine: www.msn.com
Key Search Word: lithosphere
Review: This site gives examples of how the motion of the lithosphere can affect the crust. The site is maintained by the University Corporation for Atmospheric Research. This site tells the learner that there are many surface-feature clues that let us know the lithosphere is sliding. "Lithosphere" is defined as the crust of the Earth, which consists of many tectonic plates. The site explains that two types of features form when plates move apart. These features are midocean ridges and continental rifts. The site also explains that huge mountains and volcanoes can move toward each other when the plates move toward each other. The page on this site is short, and it is easy to read and has links to many pages. The page gives an excellent picture of the Earth's lithosphere. The learner can click on the picture for a full-size version of the picture. The neatest feature of this site is that it is broken into three levels. Learner cans click on beginner, intermediate, or advanced to view information that is appropriate for their knowledge of the lithosphere. The intermediate and advanced levels give the learner more details and a more sophisticated vocabulary. This site would be helpful to use when introducing the effects of the lithosphere sliding. The user would need some prior knowledge about the lithosphere.

Title: Earth Science Explorer
URL: www.cotf.edu/ete/modules/msese/earthsysflr/spheres.html
Grade Level: Elementary through middle school
Best Search Engine: www.metacrawler.com
Key Search Word: lithosphere
Review: This site was found by going through the ETE (Exploring the Environment) Web site. It is child friendly, such that it begins with a castle where the students may enter to come upon an elevator with a dinosaur as a tour guide. If you continue to the "Earth floor" you will find a blueprint of the floor. The floor is mapped out into rooms like

Diversity, Adaptation, Plate Tectonics, Cycles, Spheres, Biomes, and Geologic Time. Go to the spheres room, which is where you will learn about the lithosphere. The text is juvenile, but extremely fun to read. It defines the lithosphere basically and gives a little bit of information of the core and the mantle of the Earth. Then the information continues with plate tectonics and geologic time. This site also defines the other "spheres" like hydrosphere and biosphere. This is a site for teachers as well (although teachers must have a code to enter the Teacher's Lounge). There is also a Resource Center on each floor listing other sites that you may want to visit. All in all, a fun, user-friendly site.

Title: Geological Society—Teaching Resources-Crust and Lithosphere
URL: http://geolsoc.org.uk/template.cfm?name=lithosphere
Grade Level: College and beyond
Best Search Engine: www.yahoo.com
Key Search Word: lithosphere
Review: This site is best used for those who are preparing to teach about lithospheres, not those just starting out learning and researching them. This Web site begins with an introduction and leads into a discussion about the differences between the crust and the lithosphere. Here the site also talks about what both the crust and the lithosphere include. It ends with evidence of the lithosphere dating back to the 1960s through the present. Again, this site is best for educators. It is dry with no visuals or extras. This site is too difficult for those just beginning their studies in this field. I would not recommend this site to help anyone get a better understanding of lithospheres.

Title: The Earth's Crust, Lithosphere, and Asthenosphere
URL: http://www.windows.umich.edu/earth/interior/earths_crust.html
Grade Level: Upper elementary through high school
Best Search Engine: www.metacrawler.com
Key Search Word: lithosphere
Review: The source of this Web site is Windows to the Universe, http:www.windows.ucar.edu, at the University Corporation for Atmospheric Research (UCAR). The opening page contains definitions for the Earth's crust, lithosphere and asthenosphere. A picture on the left side shows the lithosphere on top of the asthenosphere. Clicking on it provides a full-size version. At the top of the page are three headings: beginner, intermediate, advanced. The opening page is the beginner level. Its information is simple and the font is large. Clicking on the intermediate and advanced headings links you to

other pages with more detailed information. Below the definitions are two more links to other topics about the Earth's crust. Again each of these topics has a beginner, intermediate, and advanced level with links to more topics. At the very bottom of the home page there are links to information about the sun, the planets, comets, and asteroids. Clicking on any one leads you to a wealth of information, including pictures, myths, discoveries, and games. Below the links to the planets is another set of 24 links that allow you to read answers to questions people have asked and to submit questions. You can find art books and film related to the planets and stars. There is a link just for kids and just for teachers and too many more to mention in this short space. This site is definitely worth spending time exploring for information and just for fun. It has a wealth of information, lots of fun things, great visuals and it's easy to navigate. A great site for teachers and kids and anyone else intrigued by the planets and stars.

Title: Every Place Has Its Faults
URL: http://www.tinynet.com/faults.html
Grade Level: Grade 6 and up
Best Search Engine: www.metacrawler.com
Key Search Words: lithosphere education
Review: This site is short and sweet. It describes four types of faults with detailed explanations and moving graphics to support their descriptions. This site would provide only for a day or two of instruction, but if you wanted to make the four types of faults clear, then this site would do the job. There is also a section of good links at the end of the page.

Title: The Earth's Sphere
URL: http://www.citytel.net/PRSS/depts/geog12/litho
Best Search Engine: www.yahoo.com
Grade Level: Middle school through college
Key Search Word: lithosphere
Review: This Web site offers a huge amount of information focusing on the various aspects of structure, composition, formation, and dynamics of the lithosphere. There are 10 links on the main page: The Structure of the Planet, Plate Tectonics, Earthquakes and Volcanoes, Diastrophism, Weathering & Erosion, Rivers and their Landscapes, Glaciation and its Landscapes, Desert Landscapes, Ground Water and Landscapes, and Coastal Landscapes. The Structure of the Planet features a definition of lithosphere and a link to pictures showing the interior structure of the Earth, as well as detailed information about

the internal structure of the Earth. Plate Tectonics features maps which illustrate the planet's tectonic plates. Within the description, keywords can be clicked on to get more information and pictures. Earthquakes/Volcanoes provides information and similar links. Diastrophism offers a definition, along with two links, Folding and Faulting. These links provide more detailed definitions and pictures of these terms. Glaciation provides information about glaciers, maps, and links to information about alpine and continental glaciation. Desert Landscapes provides a definition of deserts and provides four links to the different landforms and landscapes: alluvial fan, arroyo, bajada, and sand dunes. This site is an excellent source of information about the lithosphere. The information is presented in a clear, concise manner. The information is appropriate for students in middle school through high school. I was, however, unable to connect to the following links as 404 Errors appeared when I clicked on each: Weathering and Erosion, Ground Water and Landscapes, and Coastal Landscapes.

Title: Lithosphere Mantle Interaction
URL: www.agcrc.csiro.au/project/3034mo
Grade Level: College
Best Search Engine: www.hotbot.com
Key Search Word: Lithosphere
Review: This Web site is a summary of a project conducted by Dr. Greg Houseman of Monash University. This site has six links that explain the research and give key points for doing the research. The first link gives a research summary, the second gives the goals for the project, the third explains the research strategy, the fourth gives the research results, the fifth lists related sites for further research on this project, and the sixth gives publications relating to this project. This site's opening page shows the Earth and describes how the convection in the mantle and deformation in the lithosphere occur. I think this site is good for the person who is studying the lithosphere and wishes to view research done in this area.

Mapping

Title: NIMA Home
URL: http://164.214.2.59/nimahome.html
Grade Level: Professional, specialist, field expert
Best Search Engine: www.yahoo.com

Key Search Word: mapping

Review: NIMA stands for National Imagery and Mapping Agency. There is a section on Geospatial Intelligence. It states that this is a "US Government computer system," advising visitors to read the disclaimer before continuing. Under this link you will find sections called: Earth—info site; a map and imagery server (which provides access to international maps); a digital nautical chart home page; the Geospatial Sciences Division home page; the aeronautical home page; a home page for geospatial standards; and software tools access sites. There is also a computerized imagery library access link. (I will try this one—I am a little nervous going into anything that is governmental, but I am curious about the libraries.) It takes you to the U.S. Department of Defense.

Title: Mapping

URL: http://www.geotools.sourceforge.net

Grade Level: Professionals, managers, technical staff, network administrators

Best Search Engine: www.altavista.com

Key Search Words: geographical mapping, mapping

Review: This mapping Web site is an open source site that consists of two projects, Geotools I and Geotools II. An open source project lets various sources post information regarding the topic on this site. The purpose is to provide resources for clients who are interested in interactive geographic visualization. The emphasis is on client-centered execution. Geotools II is also an open source project. It refers to using a Java GIS toolkit for developing open GIS-compliant solutions. It has modular architecture, which allows extra functionality to be added or removed easily. It lists as its long-term goal being to promote use of its core Automated Processes, Inc. (API) to become a recognized and standard API for geospatial development. This Web site is maintained by SourceForge and updated daily. It even lists the time of the update. Sections of the Web site include current and previous chat logs, downloads, members list, sections, and links. The download links go to examples and demos, applets and applications, external projects, development snapshots, and releases. Each has its own home page and a section to report any broken links. It took my computer too long to open these demo programs, over 15 minutes. Those in this industry may be familiar with open source projects and/or SourceForge, however, no webmaster or author is referenced or credentialed. Any e-mails are submitted to SourceForge. Overall, Geotools recommends software or memory expansion software in order to run various mapping tools. The same information through Source-Forge is available in several languages. Unless one is familiar with this level of geospatial development, it seems to be an extremely technical Web site.

Title: Census Bureau Geography
URL: http://www.census.gov/geo/www/index.html
Grade Level: High school through college
Best Search Engine: www.metacrawler.com
Key Search Word: mapping
Review: This Web site was a little hard to navigate at first. It is maintained
    by the government and houses information gathered during the 2000
    census. The Web site has many great features. There is an area where
    you can search for information on a particular state or town. This
    search options allows you to gather data on areas around the country. I
    feel that this would be helpful for students who are completing a class
    project. In addition teachers can also access this site for information
    when planning a lesson. This site allows you to access both local and
    national maps. Students could access maps to study different areas of
    the country. The site it even has a reference section complete with a
    glossary of terms to assist users who are searching for information.
    There is a wealth of information provided by this site. It is well main-
    tained and offers many helpful features.

Title: National Geographic Xpeditions
URL: http://www.national.geographic.com/xpeditions/standards/
    01/index.html
Grade Level: K–12
Best Search Engine: www.metacrawler.com
Key Search Words: geography & mapping & education
Review: This Web site was developed by National Geographic for teacher
    reference and use. This site divided into four sections: Activities,
    Standards, Xpedition Hall, and Lesson Plans. Xpedition Hall is an in-
    teractive area that allows a student to go on a journey that investi-
    gates geographic sites and discover various land formations.
    Standards provides information on standards for teaching geography,
    including mapping. These standards are linked to activities and les-
    son plans. Both of these options enable you to find activities and les-
    son plans that will address the appropriate standard. For teachers
    interested in teaching mapping, they can go to the standards page

and determine which standard addresses mapping (standard 1 and 2). Then switching to the lesson plan option will provide grade appropriate lesson plans to teach these standards. Additional activities for standard 1 and 2 can be found in the activities option. In addition to these four features at this Web site, there is an Xtras feature that provides additional information, activities, printable maps, interactive activities, and additional basics on mapping. This was one of the few Web sites available to address teaching mapping in the classroom. It was problem-free and covered a wide range of grade levels. The lesson plans were well explained and doable in the classroom. Additional mapping resources are available at www.nationalgeographic.com.

Title: Mapping
URL: www.cfm.ohio-state.edu
Grade Level: Grades 9–12
Best Search Engine: www.google.com
Key Search Words: mapping
Review: This is the site for the Center of Mapping at The Ohio State University. The site's introduction says: "The Center is both a NASA Commercial Space Center and an Ohio State University interdisciplinary research center focused on special data technologies, including remote sensing geodesy using the Global Positioning System (GPS), inertial navigation systems (INS), photogrammetry image processing, computer vision, image understanding, modeling and Geographic Information Systems (GIS)." There are many sections, easily navigable from a links bar. The most interesting section is one on research into developing different media for mapping. The researchers are developing a method of detecting undetonated explosives from the air, for the military. Other projects include an Airborne Integrated Mapping System, a Real-time GPS System, and an ITS Curve Warning System. There are links for publications, job opportunities, workshops, and other mapping sites. I find this site to be informational for problem-driven research programs.

Title: USGS: Science for a Changing World
URL: http://mapping.usgs.gov
Grade Level: Kindergarten to adult
Best Search Engine: www.metacrawler.com
Key Search Word: mapping
Review: The Web site is divided into two sections. Mapping News offers products and interactive CDs that you can purchase online. The sec-

ond section is Mapping Products and Services. This is the more interesting of the two. It includes interactive information, information on national spatial data infrastructures, product retailers, general information, prices, and ordering. It also includes partnership opportunities and a mapping library and publications and finally a section for parents, teachers, and students. The general information link allows the user to view maps and aerial photos. This was fun. It starts with a simple map of the United States. The user clicks on a desired area until reaching the specific location. The section for parents, teachers, and students was my favorite. Students can obtain help with homework and also get ideas for related projects. "Explorers" is a section where resources can be found for anyone who is curious about natural science. The section for teachers was full of lesson plans and activities from kindergarten through high school. There was also a section called "fun stuff" where kids (or adults) can download pages to color, bookmarks, games, and other fun activities related to natural science. This Web site is particularly good since it offers something for everyone.

Title: Total Ozone Mapping Spectrometer (TOMS)
URL: http://jwocky.gsfc.nasa.gov
Grade Level: Secondary and college, possibly honors-level middle school
Best Search Engine: www.google.com
Key Search Word: mapping
Review: This Web site is managed by NASA and is technically considered a U.S. Government Public Information Exchange Resource. It is updated regularly. Information about several spacecraft can be found: Adeos, Earth Probe, Meteor 3, Nimbus 7, Triana, and Omi. There are also topics listed that affect the ozone layer, those being Aerosols, Ozone, Reflectivity, and Erythemal UV. Perhaps the thing I liked most about this Web site is the last group of topics offered, which included links to related sites, a TOMS News section, and most importantly a section titled Teachers. Here teachers are able to explore ideas for the classroom. This site also includes ozone maps and data. Through this site one can monitor the current Southern Hemisphere Ozone Hole. Overall, this is a great government-run site especially helpful to education.

Title: Your Sky
URL: http://www.fourmilab.to/yoursky
Grade Level: Middle school adult
Best Search Engine: dogpile.com

Key Search Word: mapping

Review: Your Sky allows the user to make an actual sky map. This is of great interest to me. Which is why I chose this site. The user can make maps for any time, date, viewpoint, or observing location. The site will compute current location of orbital elements of an asteroid or comet. There is a control panel that allows you to plot, color, scheme, size the image and place parameters on your objects. The first is the sky map. The user can plot the latitude and longitude of their star or planet or select a nearby city. Horizon Views show the stars above the horizon from a specific date and time. The Virtual Telescope is the Hubble Space Telescope of the web. With the control settings you can track asteroids and comets by again setting time and an aiming point. You can even preset the controls to your own preferences. The site offers related software and resource sites, that include use of the Web, Windows, and Unix. The three maps on the Web site are clear, but the telescope map is the easiest to read.

Title: Mid-Continent Mapping Center
URL: http://mcmcweb.er.usgs.gov/index.html
Grade Level: Grade 8 and up
Best Search Engine: www.metacrawler.com
Key Search Word: mapping

Review: This physical geography Web site is designed for children in grade 8 and up to explore. This seems to be an advanced Web site to gain the further knowledge of mapping. The site was created by the U.S. Geological Survey. The site is updated frequently and encloses a privacy statement and disclaimer for validity. This Web site offers projects, products, and much more. The Mid Continent Mapping Center is one of several mapping centers and Web site offered with the assistance from the USGS. The main page has an explanation of the center, goals, and history of the company. Its main page offers many icons to explore for those interested in using systems, viewers, and appropriate topographical maps. This Web site is an advanced Web site for those who are interested in further exploration. The main page offers an icon called the Learning Web. The Learning Web is a useful tool for students, teachers, and explorers. Students can use the resources and get concepts they may need for their homework. Teachers can use this tool to gain more ideas and lesson plans for their classrooms. Explorers can use this tool to gain more knowledge of any special topics. Furthermore, this Web site overall is beneficial for all students and educators.

Meteorology

Title: The Bureau of Meteorology
URL: http://www.bom.gov.au
Grade Level: Grades 9–12
Best Search Engine: www.metacrawler.com
Key Search Word: meteorology
Review: This Web site is a great resource for teachers and students alike. There are many colorful maps for the researcher to view and study. The setup of the page is clear and understandable. The material can be viewed by younger grades, but they would have to have a parent there to help them understand the navigation and vocabulary. Overall I would recommend this site to any researcher.

Title: The Online Guides Meteorology
URL: ww2010.atmos.uiuc.edu/%28gh%29/guides/mtr/home.rxml
Grade Level: High school to undergraduate level
Best Search Engine: www.dogpile.com
Key Search Word: Meteorology
Review: As you move into the site you come across pictures of maps, tornadoes, clouds, and satellite images. On the left hand side there is a column with the words WW2010 University of Illinois. Under this heading there are three sections: online guides, meteorology and WW2010. Links in each section give you information on such topics as projects, activities, and hurricanes, hydrologic cycle light, and archives. The middle part of the page contains pictures and titles. There is a paragraph that talks about the online guide of meteorology and the modules available to the browser. There are lights and optics, air masses, weather forecasting, severe storms, forces and winds, clouds and precipitation, hurricanes, El Niño, and hydrologic cycle. As you navigate through the modules you will come across pictures and information on that topic. Each subtopic in the module gives other links and information. Most of the sites were last updated in the late 1990s. Graduate students and faculty, working through the collaborative visualization project that was funded by the National Science Foundation, developed the site.

Title: Science Clips: Science and Technology Graphics
URL: http://scienceclips.com
Grade Level: All grade levels
Best Search Engine: www.metacrawler.com

Key Search Words: meteorology education

Review: This Web site is full of clip art, both still and animated. It is excellent for students (or teachers) who are preparing visual presentations like Power Point or even creating their own Web site. Although I discovered this site through its meteorology clip art, it's full of clip art from all aspects of science such as geology, astronomy, and oceanography. There is nothing to learn here, but it's still a very good resource for creating an effective science lesson.

Title: Bureau of Meteorology (Australia)
URL: http://www.bom.gov.au
Grade Level: All grade levels
Best Search Engine: www.metacrawler.com
Key Search Word: meteorology
Review: This site was awesome! One that I would definitely use for my third-grade science class. Although it mainly used Australia for research studies, it contained many interesting links for the United States. First there were links to other sites where one could learn about meteorology and scientific terms. There was a link to educational findings as well as publications and news about the bureau. However, the most interesting to me was the curriculum materials. This link contained lesson plans for different age levels, topics in meteorology, and such. And of course, this site also had an update log.

Title: Weather: What Forces Affect Our Weather?
URL: http://www.learner.org/exhibits/weather
Grade Level: Grade 8 and up
Best Search Engine: www.yahoo.com
Key Search Word: meteorology
Review: This is a nice Web site maintained by Annenberg/CPB (The Corporation for Public Broadcasting) with links to other sites that serve as resources for teachers who use their programming in class. The Web site is almost entirely text. It includes pages on the atmosphere, water cycle, powerful storms, ice and snow, forecasting, and our changing climate. It also has a nice links page for students wishing to explore the topic further. Each of the topic pages includes a question text link. These links lead to pages answering the question. These questions include: Is our atmosphere threatened by human activity? How can ocean cycles effect the weather? How do hurricanes form? Are ice ages cyclical? Should we expect another one soon? What tools do forecasters use? There is also a nice "stormchaser" ac-

tivity. This activity requires students to plot the course of a tornado. Overall the site is excellent in its content though lacking in illustrations.

Title: Bad Meteorology
URL: http://www.ems.psu/~frazer/BadMeteorology.html
Grade Level: Middle school teachers
Best Search Engine: www.metacrawler.com
Key Search Word: meteorology
Review: This Web site was prepared by Alistair B. Frazier of the Department of Meteorology in the College of Earth and Mineral Sciences of the Pennsylvania State University. The home page opens with a dedication to "teachers who wish to get it right." According to Mr. Frazier, many well-understood phenomena in science are presented incorrectly and it is the purpose of this Web site to correct these misrepresentations. The home page begins with Bad Meteorology and as you scroll down there are links to examples of Bad Meteorology. As you click on each link you are taken to a detailed explanation of the currently accepted explanation of the phenomena and the real scientific explanation. At the bottom of these pages are links to Frequently Asked Questions (FAQ). Clicking on this link gives further detailed information. There are also links to Bad Astronomy and Bad Chemistry. The site is easy to navigate with easy access to all links. Some of the information was a little hard for me to follow but a science teacher would probably find it interesting. The tone of site is a bit preachy but its worth looking at.

Title: Students & Teachers, Bureau of Meteorology Australia
URL: http://www.bom.gov.au/lam/Students_Teachers/learnact.htm
Grade Level: Teachers for grades 3–8
Best Search Engine: www.google.com
Key Search Word: meteorology
Review: This is an excellent site for teachers to use with students. The Web page is maintained by the Bureau of Meteorology Australia. The section I reviewed is titled Students and Teachers. The site is five pages long and provides all contact info for the curriculum person, phone, e-mail, and address. The site is broken into five areas with lesson plans provided for each domain. The lesson plans are for grades 2/3–7/8. Science has 21 lessons. Math has 4 lessons, Health PE 1 lesson, Geology 1 lesson, and the arts 1 lesson. The lessons contained on the site are reproducible. The site is updated regularly. No special

software is needed to navigate the site. Again this is a great Web site for educators in these grade levels.

Title: Dan's Wild Wild Weather Page
URL: http://www.wildwildweather.com
Grade Level: All grades
Best Search Engine: www.yahoo.com
Key Search Word: meteorology
Review: This is a great site for those just beginning their studies in meteorology. This site was put together by Dan Satterfield, the chief meteorologist for a news channel in Alabama. It is broken down into every possible topic you could think of: radar, tornadoes, clouds, precipitation, lightning, humidity, satellites, temperature, forecasting, hurricanes, wind, and climate. Each category gives great definitions, pictures, and many learning activities and projects. This site also includes games, puzzles, and quizzes. This is also a great resource for teachers when planning activities for the classroom. This Web site provides information on anything and everything you ever wanted or needed to know. I highly recommend this site for anyone studying meteorology.

Title: Weather
URL: http://www.noaa.gov
Grade Level: All grade levels
Best Search Engine: www.google.com
Key Search Words: meteorology/weather
Review: This site is brought to you by the National Oceanic and Atmospheric Administration (NOAA) and gives the latest information on the weather. This site contains the most up-to-date satellite images, weather radar, and surface forecast. It includes links to weather warning sites, as well as weather archives including 107 years of weather data and aviation, space, and marine weather. This site is useful to all people in all areas. You can use this site to see a three-day weather forecast of your own area! There are links on the right side of the site to other hot topics, products, news, and organizations. All in all, I was extremely interested in surfing this site and have added it to my "favorites."

Title: Meteorology
URL: ww2010.atoms.uiuc.edu
Grade Level: Grade 8 and up

Best Search Engine: www.dogpile.com
Key Search Word: meteorology
Review: This online meteorology guide is a wonderful, user-friendly site
    that uses color and fabulous pictures. It contains a collection of Web-
    based instructional modules that use multimedia, technology, com-
    puter simulations, audio, and visual to introduce concepts. This site
    contains links for classroom activities. Another link includes remote
    sensing, reading maps, projects, and activities. The Meteorology sec-
    tion includes subjects from air masses through weather forecasting.
    This site is great and one to see.

Title: Meteorology Online
URL: http://library.thinkquest.org/C0112425
Grade Level: Elementary through high school
Best Search Engine: www.yahooligans.com
Key Search Word: meteorology
Review: This is a student-created Web site. The Thinkquest site offers
    Chinese, English, and French versions. The site offers students an in-
    teresting and interactive learning experience related to meteorology.
    The information ranges from basic to advanced. The creators of the
    site advise users to visit the Information page for help navigating the
    site. On the information page, you will find detailed descriptions/
    information about the different pages and their setup. Here you can
    learn the functions of the various bars and buttons displayed on the
    site. On the main page, there are links to the five sections: Children,
    Student, Library, Laboratory, and Staff Office. Children: This section
    of the site has four links, Chapter, Resources, Courses, and Games.
    The Chapter section features information about the different
    elements of weather. The pictures and diagrams are very clear. Re-
    sources: This section provides links/information for additional infor-
    mation about meteorology. Courses: In this section, more advanced
    topics are covered, such as El Niño. Games: A quiz is given to test
    your knowledge. Student: Detailed information is provided about
    sun, wind, storms, rain, and so on. Users can easily navigate to the
    different topics by using the backward/forward buttons at the bottom
    of the page. Library: This section provides even more information/
    pictures and links to the information that is covered in the Chapter
    sections found in the Children and Student sites. Laboratory: Stu-
    dents can do experiments that are guided by flash movies. Staff Of-
    fice: This section provides profiles/pictures of the students who
    created the Web site. I would definitely recommend this site. It has a

wealth of information that is suitable for students elementary through high school. There are plenty of pictures/diagrams and the terminology is easy to understand.

Methods for Uncovering Earth History

Title: Chapter 3: History of Life on Earth
URL: http://www.idiotsguides.com/Chapters/0028631994_CIG_Life_Science/file.htm
Grade Level: All grade levels
Best Search Engine: www.yahoo.com
Key Search Words: earth science, uncovering earth history
Review: This is the Complete Idiot's Guide version, chapter 3. It covers the following topics: 1. Macroevolution, 2. The fossil record, 3. Measuring geologic and evolutionary time, and 4. Mass extinction of the dinosaurs. This appears to be the actual chapter from the book. It is simplistic in its discussion. It contains different little sections that help certain parts stand out. (They use a cute little globe, with faces.) The sections are Bio Buzz, Think About It, and Try It Yourself. In Bio Buzz, they define in simple terms such concepts as evolution. The Think About It section directs your attention to other sources such as other readings, and even suggested watching *Jurassic Park*. The Try It Yourself portion suggests making a fossil from plaster of paris to get an idea of what a fossil looks like. At the end of the chapter, there is a section called The Least You Need To Know, which gives a few tidbit facts. I liked this page because I feel geologically impaired; I really don't know too much about Earth Science, and this material would help me learn all the things I missed in school, plus it was rather enjoyable to read!

Title: The Deep Earth
URL: http://www.ras.org.uk/pdfs/g_uk/pp2425.pdf
Grade Level: College and professional
Best Search Engines: www.yahoo.com; www.metacrawler.com
Key Search Words: earth's history, uncovering earth's history, seismology
Review: This Web site's title is The Deep Earth. It is about uncovering the deep mantle and its application to seismology. The article briefly addresses the history of the first steps in discovering the structure of deep earth and the naming of the study as seismology. The article goes on to discuss the application the science of seismology has had on determining the Earth's subsurface structures down to the core. The credibility of this information, which comes out of the United

Kingdom, is discussed. The theory of the development of the science is revealed, along with some of the tools seismologists use. As one theory is hypothesized, tested, and proven through analysis, expansion of the science occurs and these implications are explained. A new approach to understanding the composition of the deep earth is discussed and its related sciences. The article, although compact, discusses terminology and definitions. It ends with even more specific applications of geodynamic operation and Earth's magnetic field. Unanswered questions about the core mantle are postulated. Other than *Geophysics in the U.K.*, being listed at the top of the article, the author of the article was not given. Further investigation into the Web site revealed that the article comes out of the Royal Astronomical Society, the United Kingdom's leading body in astronomy, geophysics, and planetary science. It was established in 1820 and has amassed a vast library and astronomical instruments. The publications the society is credited with come from Blackwell Science Limited and include *Monthly Notices of the RAS*, *Geophysical Journal International*, *Astronomy & Physics*, and reviews for *Astronomy in the UK* and *Geophysics in the UK*. Author guidelines for publication are also included. This Web site resource appears to be highly credible.

Title: Geologic Time: The Cenozoic
URL: http://www.mnh.si.edu/antro/humanorigins/faq/gt/cenozoic/
      cenozoic.htm
Grade Level: Middle and high school
Best Search Engine: www.yahoo.com
Key Search Words: earth science & earth history
Review: This Web site was located through the Museum of Natural History's Web site. This site provides a great deal of information about the Cenozoic era, which is Earth's current geologic era. The opening page of the site provides a timeline that provides a visual depiction of the Cenozoic era and how it is broken into the Tertiary and Quaternary periods. The timeline is further delineated to depict the five epochs of the Tertiary era (Paleocene, Eocene, Oligocene, Miocene, and Pliocene) and the two epochs of the Quaternary period (Pleistocene and Holocene). The site provides an overview of the Cenozoic era and provides links to each of the eras and corresponding epochs. In addition to providing information on each of the eras and epochs, a Geologic Timescale is provided in visual form with written explanation. This timescale shows the Phanerozoic eon, which is the current eon, broken down into eras. This timescale assists the user in

having a better understanding where the Cenozoic era falls within Earth's total history. This site provides middle school and high school students with a great deal of information and will enable them to glean a higher level of understanding of these eons, eras, and epochs in Earth's history. This site provided information in visual form, included some pictures, and was problem-free to navigate.

Title: Methods for Uncovering Earth's History
URL: http://seaborg.nmu.edu/earth
Grade Level: Grades 9–12
Best Search Engine: www.google.com
Key Search Words: earth history
Review: Earth History Resources is like a textbook on the Internet. The main page gives a great overview and explains how to surf the site. "The illustrations and pictures throughout the site are able to be copied and downloaded for educational purposes." The five major links off of the main page are: Life through Time, Geologic Timelines, Mammoth Site, Calendar of Time, and The Museum. Life through Time is a pictorial history of life on earth. It is accompanied by a narrative of that period. The first era is the Cenozoic and the first period in the era is the Quaternary. The first epoch in that period is the Holocene, which was 0.01 million years ago, otherwise known as the end of the Ice Age. This is just an example of how descriptive a site this is. A geological timeline is used to simplify the history of Earth by shrinking it down to a half-day, birth being at midnight and the present being noon. Each major event in time winds up being every hour through that half-day. This is a great site for high school students to use as an additional source of information.

Title: The Time Machine
URL: http://www.uky.edu/ArtsSciences/Geology/webdogs/time/
    time4.htm
Grade Level: 5 and up
Best Search Engine: www.google.com
Key Search Words: earth history
Review: The Time Machine is a link on the Web DoGS site (http://www. uky.edu/ArtsSciences/Geology/webdogs/welcome.html), which is a resource page covering various concepts related to Earth history. On the Time Machine page, students can click on a period in the Earth's history and a page gives a brief summary of the period along with a graphic image. The Time Machine is easy to use, and it would have

many applications to classroom instruction. The entire Web DoGS site contains much information on earth history. From the home page of this site, there are links to geology and the Civil War:

The Amber Page

Tom's Stream Table Page (not a working link)

Java Page—Simulated earthquake where students click on a building to begin the earthquake process

Ask a Geologist

Virtual Plates—Simulated plate tectonics with a link to see how fast the plate you are on is moving

Rocks on the Web

Kentucky Earthquake Page

Quick Minerals

Time Machine

These links are listed in a vertical fashion and must be accessed from the home page of the Web site. While one of the mission statements of Web DoGS is to have a site that is easy to navigate, it has not succeeded in doing this. In most cases you have to use your browser's Back button to return to the home page. In a few cases, the links did not work, or the page was no longer available. However, the information available on the site is worth viewing.

Title: Paleomap Project
URL: http://www.scotese.comdefault.htm
Grade Level: Middle school through high school, and teachers
Best Search Engine: www.metacrawler.com
Key Search Words: history of the earth
Review: Paleomap Project was developed by Christopher R. Scotese. The user is able to travel through time to view the world from the distant past and from far into the future. You can click on "more information" to read detailed data about each particular time period. The site includes sections on several areas. The site map is an index of all of the links/information available on the site. The site also includes a section on climate history, current research projects by Dr. Scotese, paleomap software, and order forms. There is also a link labeled "animation." This takes a while to download. To tell you the truth, even after I read the attached description, I still wasn't too sure about what I was viewing! Overall, the site is fun and user friendly. I recommend it to anyone who may be doing research on the history of the Earth

and also for those who already have an understanding of the subject. Science teachers may find it to be a useful resource.

Title: The Earth's History through Pictures
URL: http://gallery.in-tch.com/earthhistory
Grade Level: Elementary and above
Best Search Engine: www.dogpile.com
Key Search Words: earth's history
Review: This sites includes info about the author and illustrator, who is one in the same, Douglas Henderson. The site begins with the geologic timeline, which represents the time on Earth. The author begins at the Quaternary period (which is now) and goes back to the Archeozoic era, when the earth's crust was formed. The author uses several beautiful illustrations to show all of the changes. It takes a while for the images to download. I found some of the words a bit difficult for elementary students. This is a good site if one wanted to find the different eras. It is good if one is beginning to study the Earth's history.

Title: Geophysics in the U.K.
URL: http://www.ras.org.uk
Grade Level: High school
Best Search Engine: www.webcrawler.com
Key Search Words: methods for uncovering earth history
Review: This British Web site provides both text and a visual depiction of what it refers to as "this vast inaccessible region," the mantle of the Earth. This site discusses the importance of the use of seismology in the study of the Earth's history. It discusses topics such as seismic stations, topographic analyses, and plate movements. There is a brief discussion of the early scientists involved in this field. This geophysics site would serve as one research source for a high school student studying this topic. It does not, however, offers any links.

Title: DIG: The Archaeology Magazine for Kids
URL: http://www.digonsite.com
Grade Level: Grades 3–8
Best Search Engine: www.metacrawler.com
Key Search Word: archaeology
Review: This physical geography Web site is designed for children 8–13 years old. This Web site is presented in an easy and fun to read man-

ner. It is published by the Cobblestone Publishing Company, a division of the Cricket Magazine Group, in cooperation with the Archaeological Institute of America. This Web site offers the current issue of *DIG: Archaeology Magazine for Kids*. There are icons to explore on the left side of the main page. They are labeled DIG, home, subscribe, quiz, ask dr. dig, fantastic factoids, glossary, links, parents and teachers, and about us. I explored all of them and they all seem to be user friendly. There are interesting facts and a state-by-state guide to archaeology and paleontology events for kids, families, and schools. This Web site is suitable for children of ages 8–13 who want to explore archaeology, paleontology, and Earth Science. I recommend this site to every teacher, parent, and child since it's user friendly and fun.

Minerals

Title: Department of Minerals/Energy
URL: http://www.dme.gov.za
Grade Level: College
Best Search Engine: www.google.com
Key Search Word: minerals
Review: As you click to the Web site you come across the heading and five pictures. On the right side under the heading is the Republic of South Africa. On the left side you have nine items to choose from: About the DME, Energy, Mineral Development, Mine Health and Safety, Publication, What's New, Links, Happenings, and News Center. As you navigate through the nine items you will come across information concerning South Africa's development through minerals and energy. One of the nine items will give you the mission statement, contact list, the ministers and services. If you click onto the Happenings site it provides you with information on events, exhibitions, and meeting. This site talks about the mineral development. Mine Safety talks about accidents and the statistics. It discusses examinations, mine surveying, mine equipment, and the Mine and Health Act. The links part contains sites for mineral economics, mineral development, energy, and general information. The site allows you to perform a search by typing in your information. The site is informative and is geared toward people who want to find out about South Africa. It does not contain a lot of information for you if you are a beginner student.

Title: Franklin & Sterling Hill Minerals
URL: http://simplethinking.com/franklinminerals/franklin.stm
Grade Level: High school
Best Search Engine: www.metacrawler.com
Key Search Words: minerals education
Review: This site is dedicated to the Franklin-Sterling District in the
    northwest corner of New Jersey. Franklin-Sterling is home to 340 dif-
    ferent minerals, of which 80 are fluorescent—the most anywhere in
    the world. This site focuses on a particular type of mineral, so it is not
    good for a general lesson plan on minerals. What was most appealing
    about this site was its fluorescent mineral gallery. It contained a
    photo gallery of several fluorescent minerals, complete with descrip-
    tions. If you are doing a unit on minerals, it is worth checking out.
    The site also contains a general mineral section, but like I men-
    tioned, only on a particular type, so it's very specific. The site's worth
    lies in its entertainment value, as some of the photos are really great.

Title: Geology
URL: http://geology.about.com
Grade Level: College
Best Search Engine: www.metacrawler.com
Key Search Word: minerals
Review: This topic was difficult to research because "minerals" is a syn-
    onym for "vitamins." There were hundreds of solicitations for vita-
    mins. However, this site was suitable for information about minerals.
    There were pictures of every mineral you could possibly imagine,
    which were cool to look at. This site has a few links to other informa-
    tional areas about minerals and their location. I didn't see an update
    name/number or how many times this site was updated.

Title: The Mineral Identification Site
URL: http://www.netspace.net.au/~mwoolley/top.htm
Grade Level: Grades 4–10
Best Search Engine: www.yahooligans.com
Key Search Word: minerals
Review: Since the site is being updated, a limited number of links are in full
    working order. From what I have seen I can say with confidence that
    this is a useful site. There is a lot of potential for identifying advice and
    general knowledge about minerals. The site has won many awards for
    excellence. Teachers and students will benefit greatly from this site.

Title: Rocks for Kids
URL: http://www.rocksforkids.com
Grade Level: Grades 5–10
Best Search Engine: www.metacrawler.com
Key Search Words: minerals and kids
Review: This site was developed to help students learn about rocks and minerals. It gives the users great information and offers many links to other informative sites that have great pictures, stories, and information activities. Students can easily find the information they are looking for by simply clicking on the Table of Contents link. Within the Table of Contents there are 15 categories about rocks and minerals: How Rocks and Minerals Are Formed, Quarries, Identification of Rocks and Minerals, Uses of Rocks and Minerals, Hands On Experience Lab Outline, Collecting Rocks and Minerals, National Disasters, Arts & Crafts, Fossils, Teachers' Corner, Photos of Some Rocks & Minerals, Some Collecting Locations, Bert Ellison articles, Bill Kovacs projects, and Links. This site offers numerous amounts of information within each link. The information is clear and easy to understand. Most links offer pictures, charts, and graphs to help the learner understand the information presented. There is also a glossary on some pages to clarify information. Within each link, the user can link to even more sites to get more information about a specific topic. Teachers can also benefit from using this site. The Teachers' Corner link allows teachers to access quizzes, lesson plans, and other rock sites on the Web. The user can also contact the creators of this site through e-mail. I would recommend this site to anyone interested in general information about rocks and minerals.

Title: A Wonderful World of Minerals
URL: http://library.thinkquest.org/J002744/adlm.html
Grade Level: Elementary and middle school
Best Search Engine: www.yahooligans.com
Key Search Word: minerals
Review: This Web site, created by students from the John F. Pattie Elementary School in Virginia, offers 12 links and contains a wealth of information about minerals. The information is written by kids, so it is easy to read and understand and perfect for elementary level students. A diagram of Moh's Scale, which organizes rocks in the order of hardness, is featured on this page. The Birthstones link provides the name of each birthstone, along with a color picture. If you click on the name of the

birthstone, you will be taken to another page that provides details about the birthstone, such as color and where it is commonly found. The Growing Crystals link gives directions on how to grow crystals, and it also has a link to detailed instructions for a crystals science project. There is also a link to a worksheet to be used for observations. The Games link offers a word search and memory games based on rocks. The Metals link provides information about copper, titanium, zinc, chromium, mercury, nickel, iron, tin, uranium, magnesium, plutonium, and alloys. Each metal features another link that gives basic information, such as a description, atomic number and uses. It does not provide any pictures, however. The Rock Photo Album has great pictures and detailed descriptions of a variety of rocks and minerals, such as obsidian, pumice, geode, fluorite, amethyst, amazonite, pyrite, and quartz. The Bibliography link lists the books used by the student creators of the site, as well as the actual links to the electronic resources used. I would recommend this site to elementary students, as well as to middle school students who will also benefit from the information on this site, as it is a great starting point for reports or projects.

Title: James Madison University Mineral Museum
URL: http://csm.jmu.edu/minerals/default.htm
Grade Level: High school and above
Best Search Engine: www.yahoo.com
Key Search Word: minerals
Review: This site provides the visitor with photographs of a wide variety of minerals. Many of the photos are accompanied by a paragraph or two providing basic information about the mineral including how it was formed and its commercial uses. There is a section called micro mounts where visitors can look at specimens under magnification. While this Web site is a good idea I felt that the pictures were of a low resolution and as such difficult to fully appreciate. This is obviously a poor substitute for first hand contact with the minerals. However, it is a handy reference and may help students identify some samples.

Title: Crystals, Minerals, Precious Metals and Gemstones
URL: http://www.jewelrysupplier.com
Grade Level: All grade levels
Best Search Engine: www.metacrawler.com
Key Search Word: minerals
Review: This is an excellent Web site for anyone interested in minerals, gemstones, crystals, and so forth. Down the right side of the site is a list

of all gemstones that you can access for more information. Down the left side is information regarding the history of the minerals, where they can be found (geography), healing powers of the gems, properties, symbolism, mythology, and spirituality. Did you know that the amethyst is associated with peace and calming qualities? Amethyst can be used to treat toothaches, skeletal discomforts, posture and other bone and joint-related sicknesses (like arthritis). There are others if you believe in this sort of thing. It is interesting to read about, although I am not a believer. From each gem are other links to go to if you require more information. This is definitely a site worth checking out.

Title: Rocks and Minerals
URL: http://www.fi.edu/tfi/units/rocks/rock.html
Grade Level: K–12
Best Search Engine: www.google.com
Key Search Word: minerals
Review: This was a comprehensive Web site from the Franklin Institute Online. It begins by introducing rocks and geology. It then allows you to go into specific classes and categories of minerals. The category options are Mineral Gallery, Minerals by Name, Minerals from Geology Museum, Clausthal Online Mineral Collection, Alphabetical Mineral Reference, and Commercial Mineral Names. Once you select a category, it gives you more information and photos. There were also broad categories to click on. Some categories were Form Solid Mineral Deposits, Add Unusual Characteristics, Assume a Fixed and Definite Shape, and Teaching Others about Rocks. The last category I found to be the most useful as a teacher. There was a list of activities for teaching Minerals, Magmas, and Volcanic Rocks to students in grades K–12. The lessons were listed, and the suggested grade levels were also given. As a teacher, or as a parent, I thought this was extremely helpful. You were also able to search other sites, and this site also linked the user directly to other sites. Overall, I found this site worth looking into if you are interested in teaching and learning about rocks and minerals.

Title: Mineral Matters
URL: http://www.sdnhm.edu
Grade Level: Grades 4–6
Best Search Engine: www.abcsearch.com
Key Search Words: mineral teaching lesson plans
Review: Finally after browsing hundreds of sites, I found one that really works. It is great for grades 4–6. More importantly it flows well. If stu-

dents had an interest in minerals there are several projects and pieces of information they can use to continue their interest. The Web sit itself is maintained by the San Diego Natural History Museum. The site itself is broken down into five sections. 1. How to identify minerals, 2. Create a collection, 3. FAQ, 4. Grow your own crystal, and 5. Mine games. In the FAQ section the most basic definitions are given to define minerals from other rocks. In the intro nine properties of identification are introduced with links for further help and explanation. In How to create a collection, the site explains how to clean, organize, display, and store minerals.

Title: Atlas of Igneous and Metamorphic Rocks, Minerals, and Textures
URL: http://www.geosci.unc.edu/Petuna/IgMet-Atlas
Grade Level: College
Best Search Engine: www.metacrawler.com
Key Search Word: minerals
Review: This Web site was developed at the University of North Carolina as part of the Virtual Geology project. It was constructed to aid undergraduates in the Geology Department at the university. It opens with a Table of Contents, beneath which are four main headings: Index of Minerals, Plutonic microtextures, Volcanic microtextures, and Metamorphic microtextures. Clicking on the link to the Index of Minerals takes you to a list of minerals. Clicking on any of the names links you to a photographic slide with a description of what you are viewing underneath. On the right side of the slide at the bottom is an arrow that says "another example." Clicking on it sometimes takes you to another example, sometimes not. Similarly clicking on each of the other headings takes you to lists of terminology and links to photographs with an example. There are no additional links. This Web site is basically a site of photographic slides. It assumes a reasonably well-developed knowledge of the subject matter as there are no definitions or explanations of terms. For the neophyte this site might prove daunting; however, for someone well versed in the subject, it could have some merit and be worth looking at.

Title: Minerals by Name
URL: http://mineral.galleries.com/minerals/by-name.html
Grade Level: Grades 4–12
Best Search Engine: www.aol.com
Key Search Word: minerals
Review: This is an excellent site to find out anything and everything you ever wanted to know about any mineral you could think of. There are

introductory sections on the properties of minerals and on silicates. The section on properties explains how minerals are classified and the types of characteristics. Then you can search for a mineral in three ways: name, class, or groupings. You can choose a mineral by class by choosing elements, oxides, and so forth or by groupings by choosing birthstones, gemstones, and so on. If you decide to choose a mineral by name, you will get to choose from an alphabetical list. You will receive a picture of the mineral and the mineral's chemistry name, its class, subclass, group, uses, and specimens. Then you will read an in-depth description of the mineral and finally the physical characteristics. This Web site is not exciting with lots of colors and pictures, but it has a great deal of information for anyone studying minerals. It is great for younger students and for older students. I highly recommend this site.

Title: USGS
URL: http://www.minerals.usgs.gov/minerals/
Grade Level: High school and college
Best Search Engine: www.dogpile.com
Key Search Word: minerals
Review: This site dedicates itself to the person who wishes to find information in depth on minerals. The site gives information on the world-wide supply and on demand and flow of minerals and materials essential to the U.S. economy, the national security, and the protection of the environment. This site also provides statistics and information by commodity, country, and state. Furthermore, this site provides popular topics on materials flow, recycling, and historical statistics. The site includes featured publications and a section on what's new in the area of mineral research. More links include Mineral Resources in the east, west, central, and crustal imaging and characterization along with spatial data. In my opinion this site is good for mineral searches. One thing to know is that when printing or viewing certain areas a PDF reader is needed.

Oceans

Title: Welcome to Cornell Theory Center
URL: http://www.tc.cornell.edu/Services/Edu/MathSciGateway/environment.asp
Grade Level: All grade levels
Search Engine: www.yahoo.com
Key Search Words: oceans, earth science

Review: I think you will like this Web site! It has different links to visit. At the top of the page there are six boxes to choose from. One of them is Oceans, Lakes, Streams and Wetlands. When I clicked on the box, it merely forwarded me through the document to where the Web pages are listed and a brief description about the site to visit. There are 14 different links to go into. I will try one or two in which I have some interest. I enter the whale watching Web page. Okay, so I clicked into a Web site that questioned the cruelty of whales in captivity. Wrong page for me to go to. I did not like the stats I read. Back tracking . . . Wow! I found the whale and dolphin declaration of rights. It was great! Thanks, Mr. Cousteau! There is also a Jason Experiment Section that was set up after so many letters were received from children about the RMS *Titanic*. Other sections covered wetlands in America, educational resources, and oceanography from the space shuttle, earth guide, water environment federation, as well as several others. I like this site, and hope you do, too.

Title: Oceans Alive
URL: www.abc.net.au/oceans/alive.htm
Grade Level: Grades 6–12
Best Search Engine: www.google.com
Key Search Word: oceans
Review: Oceans Alive is an educational project by ABC Online, Community Biodiversity, The British Council Australia, and oz-Teach. net. This project celebrates marine diversity and exploring ways we can help save our oceans. There are eight sections off the main index. The first section is called Whale Dreams. This section is about whales and whale spotting in Australia. This section also includes interactive maps to log your own sightings. The second section, Jewels of the Sea, highlights some of Australia's top marine biodiversity areas. The section entitled Seal Training describes how seals are trained and how animals learn. Beachcomber highlights schools with online sites covering matters in marine life. The section called Sea Rangers has a question-and-answer format where a student could get an answer to any question asked having to do with marine life. There is also a collection of international links such as the Great Barrier Reef Marine Park Authority and Seaweek 98.

Title: Enchanted Learning
URL: http://www.enchantedlearning.com/coloring/oceanlife.shtml
Grade Level: Grades 2–5

Best Search Engine: Mamma (www.mamma.com)

Key Search Word: ocean

Review: This site was developed by Enchanted Learning. The site offers a wealth of information on a number of subjects and promotes the user to explore several resources about oceans and its wildlife. This Web site features 11 subcategories in the field of ocean. The page on oceans provides students with a map of the oceans and information on oceans. Students can find out vital information on ocean life and environmental conditions and offers a collection of photos, short articles, and bulletins to explore. I chose to explore a subsite called Oceans. This subsite offers an introduction to the Pacific, Atlantic, Indian, Southern, and Arctic oceans. There are interesting facts about all of Earth's oceans listed in a readable format. The site is filled with illustrations on the various animals that live in the oceans. Students can click on each animal to learn interesting facts. There is even an area to search for particular facts. This site is a colorful resource for both teachers and students. Students can use this information to create wonderful research projects. Teachers can use this site to develop interactive lessons on the wonders of the ocean. The site is well maintained and easy to use. This is a first-rate site filled with information and interesting activities on oceans. It is an excellent and fun resource for everyone.

Title: UN Atlas of the Oceans

URL: http://www.oceansatlas.com/index.jsp

Grade Level: Grade 7 and up

Best Search Engine: www.google.com

Key Search Word: oceans

Review: This Web site, the UN Atlas of the Oceans, is funded by the United Nations Foundation. The home page features four icons: Uses, About the Oceans, Issues, and Geography. Clicking on any of these icons takes you into a section with specific information about our oceans. Additionally, the left side of the screen has another menu with related pages, including new ones and recently updated pages. The About the Oceans section contains links to additional pages on various topics, some of them are: how the oceans were formed, how oceans are changing, ocean dynamics, physical and chemical properties of oceans, ocean-atmosphere interface, biology of the ocean, ecology of the ocean, coasts and coral reefs, monitoring and observing systems, research, and the ocean of the future. The uses of the ocean pages includes such topics as fisheries and aquaculture, recreation and tourism, transportation and telecommunication, energy, disposal of waste from land, and offshore oil, gas, and mining. The issues page contains various links such as emergencies, food security, economics, and safety. The geography section is sill under construction, but it does contain some active links to maps of the world's ocean. While no lesson plans are included in this site, there is a wealth of information for any classroom studying the ocean.

Title: Atlantic Ocean Geography 2000
URL: http://www.photius.com/wfb2000/countries/atlantic_ocean/
atlantic_ocean_geography.html
Grade Level: Grades 5–8
Search Engine: www.yahoo.com
Key Search Words: oceans & geography
Review: This site, which provides some basic information about the Atlantic Ocean, was provided through www.geographic.org, which is an organization that provides geographic information about the countries and oceans of the world. The site chosen about the Atlantic Ocean provides geographic information including location, map references, area, coastline, climate, terrain, elevation extremes, natural hazards, current issues related to the environment, and a geography note. The information covers the geographic extremes of the Atlantic Ocean. There is information about the economy supported by the ocean (shipping, food). When you click on Transportation you get a list of major Atlantic ports, and clicking on Transportation Issues will provide information on problems related to transportation that are unique to the Atlantic Ocean. This site provides some basic information that elementary and middle school report writers will find useful. This site provides a link to Quick Maps to download a map of the area, and if the user requires any flags there is a link to Flags of All Countries. This site was somewhat limited in the information provided, but this is sometimes preferable for younger users, who get bogged down when there is too much information. The site navigation was problem-free and provided links to information about other oceans.

Title: Oceans Alive!
URL: http://www.mos.org/oceans
Grade Level: Middle to high school
Best Search Engine: www.yahooligans.com
Key Search Word: ocean

Review: This site is done from the Museum of Science. It is broken down
   into five main topics: The Water Planet, Oceans in Motion, Life in the
   Sea, Scientist at Sea, and Resources. The Water Planet category allows
   the user to explore the physical features of the ocean, the changing
   ocean, the water cycle, and ocean profiles. From there the user can get
   even more in-depth information on a topic. When you go into Oceans
   in Motion, the related sites are the Pangaea Theory, Hydrothermal
   Vents, Submarine Volcanoes, and Hot Stuff. Looking at the Sea gives
   the user a recipe to build a model of the water cycle. The recipe was
   very easy to follow. Oceans in Motion deals with current events, winds
   and waves, and the ebb and flow of the sea. Life in the Sea breaks down
   the different ocean levels into sunlight, twilight zone, midnight zone,
   abyssal zone, and hadal zone. Scientist at Sea shows the users how to do
   remote sensing and underwater exploration. Resources has different activities and a library. I didn't like the graphics in this site.

Paleontology

Title: Site Index for USGS Paleontology
URL: http://geology.er.usgs.gov/paleo/siteindex.shtml
Grade Level: High school and above
Best Search Engine: www.earthlink.com
Key Search Word: paleontology
Review: As you click onto the site you come across the heading in the
   upper left corner. Under the heading there are 12 other links high-
   lighted in blue: The USGS home page, Fossil groups, Dinoflagellates,
   Benthic Foraminifera, Mollusks, Spores/pollen, Vertebrates, Products
   of paleontology studies, Education resources, Geologic time scale,
   Glossary of terms, and Acknowledgments. The home page gives a
   brief description of paleontology and informs you of what the following pages will show you. The Fossil group talks about the types of fossils, how they lived and answers important questions. This section is
   broken down into three parts: Benthic, Pelagic and Terrestrial. If you
   click on one of these areas it will define the words, provide pictures,
   and give other important information on these items. The Products
   of USGS gives a list of published data and information from 1990
   through 1998. The resources for paleontology gives general information and teaching sources along with general publication and
   non-USGS resources. The time scale gives a detailed version of the
   timeline. The site is easy to navigate through and provides you with a
   lot of information that is easily read.

Title: Introduction to the UCMP Exhibit Halls
URL: http://www.ucmp.berkley.edu/exhibits.html
Grade Level: High school
Best Search Engine: www.metacrawler.com
Key Search Word: paleontology
Review: This site is an extensive look into the beginning of paleontology, not just a site about dinosaurs. This site, packed with information and photos, is in depth. Any teacher who would use this site in his or her lesson plans would need to really do some homework first. The site is not easy to browse because of its magnitude. It comes with its own suggestions on how to most properly navigate through all of the information, but it would need to be checked out first. Good site for an extensive unit on paleontology.

Title: Becoming Human
URL: http://www.becominghuman.org
Grade Level: Grades 9–12
Best Search Engine: www.metacrawler.com
Key Search Word: paleontology
Review: This is a fantastic Web site for researchers and teachers. There are many resources listed alphabetically by subject. This makes further research possible and easy. The site also has many educational games.

Title: Dinosaurs
URL: http://www.dinosaur.org
Grade Level: Elementary
Best Search Engine: www.metacrawler.com
Key Search Word: paleontology
Review: This site was pretty fun. Since I used to teach dinosaurs to second-graders I found some interesting items I could use for future lessons. There were links to games, magazines, articles, and pictures. This site also linked to National Geographic sites, which provided more in-depth articles for older students. I didn't see a name or e-mail address for questions or concerns, but there was a number of how many times this site was hit and when it was last updated.

Title: Digging into Fossils
URL: http://www.isu.edu/departments/museum/education/
    homep/flc_files/subm1temp.html
Grade Level: Grades 4–6
Best Search Engine: www.metacrawler.com

Key Search Word: paleontology

Review: This site was developed by the Idaho Museum of Natural History to help kids access information about fossils and paleontology. It is a great resource for teachers who would like to teach a unit on fossils. The first page clearly defines the terms fossils and paleontologist. The important terms can be easily spotted because they are in bold red print. The second page allows the user to learn more about fossils. There are four main questions posted on this page: What is a fossil? How is a fossil made? In what rocks are fossils found? How are fossils named? What do fossils tell us? All these questions are answered in clear and easy to understand language. The site also includes colorful pictures and diagrams to help the students learn about fossils. A neat feature on this site is that there is a link where the user can download fun activities with fossils. There are fossil mazes, word finds, and puzzles. The only downside to this site is that the font color on many of the pages blends into the background. This can make it a little difficult to read some of the information. Overall, I would recommend this site to elementary school teachers who are interested in teaching about fossils.

Title: Paleontology in the News
URL: http://www.geology.ucdavis.edu/~GEL3/paleonews.html
Grade Level: Grades 6–12
Best Search Engine: www.yahoo.com
Key Search Word: paleontology
Review: This page is a frequently updated links page to breaking news in the world of paleontology. Links are to newspaper, journal, and news service Web pages with current stories about paleontology. I think this would be an excellent current events resource for the middle school or high school science teacher. The down side is that many of the links do not work. This occurs because many of the link sites remove these stories after a set period of time or require subscription to access them as "archives." This does not diminish the usefulness of the site. In this discipline where new discoveries are constantly challenging old theories, it is important to stay current.

Title: ZoomDinosaurs.com
URL: http://www.enchantedlearning.com/subjects/dinosaurs/
searchword:paleontology
Grade Level: All grade levels
Best Search Engine: www.metacrawler.com

Key Search Word: paleontology
Review: This colorful and interactive Web site caught my attention right
away. Because of its visual stimulation, it is extremely user friendly. The
table of contents is laid out in such a way that the user can go to a
whole section or skip to a specific part of the section. For instance, in
the "All about dinosaurs" section the table of contents includes subsections such as information pages about individual dinosaurs, pictures,
how dinosaurs get their names and what the names mean. In the Extinction section, the subsections include the definition of extinction,
and other theories of how dinosaurs left the Earth (for example, the asteroid theory). Other sections include fossils, anatomy and behavior,
and classification. There is a frequent questions and answers section for
all inquiring minds. This site is teacher friendly as well. There is a section of dinosaur quizzes as well as one of classroom activities.

Title: Paleontology and Fossils Resources
URL: http://www.uarizona.edu/~jount/paleont.html
Grade Level: All grade levels
Best Search Engine: www.dogpile.com
Key Search Word: paleontology
Review: The Paleontology and Fossils Resources Web page is a selected first
of Web pages dealing with paleontology fossils. The Home page con-
tains a list of topics that include the following: Associations, Clubs and
Societies, Bibliographies, Books, Careers and Employment, Catalogs,
Directories and Lists of Links, Classrooms, Courses and Teaching, Col-
lections and Collection Catalogs, Colleges and Universities, Commercial and Companies, Computers and Mathematical Modeling,
Dinosaurs, Evolution and Extinction, Fossil Collecting Guidelines,
Laws, Legislation and Policies, General Information, Geographic Regional Localities–U.S., Geographic, Regional Localities, Other Coun-
tries, Geologic Ages and Formations, Ichthyology/Trace Fossils,
Invertebrates, K–12 Level of interest, Listservs, Newsgroups and Bulletin Boards, Museum and Museum Exhibits, Paleobotany and Palynol-
ogy, Paleoclimatology and Paleoecology, Periodical Publications,
Journals and Newsletters, Research Centers, Groups, and Institutes,
Researchers and Collectors, Personal Pages, State Fossils, Taxonomy
and Systematics, Television Program, Trilobite, Vertebrates, Links to
Geosciences, Resources and Books, Library Guides, Bibliographies, and
Indexes. Each topic is followed by lists of additional links that take you
to more information on the subject. This site has to be the ultimate

Web site for paleontology. It seems like if it can't be found under one of these topics it just might not exist. The only problem I can see is that it would take many hours to investigate all the links. However, the topic headings help. If you have some idea of what you are looking for, this Web site will lead you there.

Title: Paleontology: The Big Dig
URL: http://www.ology.amnh.org/paleontology/index.html
Grade Level: Elementary and middle school
Best Search Engine: www.yahooligans.com
Key Search Word: paleontology
Review: Paleontology: The Big Dig, is the kids site of the American Museum of Natural History. It is a terrific site! The site has five main links. Going Gobi: The Hunt for Fossils in Mongolia details the adventures of the museum's scientists in Mongolia. Asterisks within the paragraphs link Web site visitors to "Ology cards," which feature a picture related to the information on the page. Information featured includes the reasons for their expedition, pictures of the places visited, a map of their travels, a typical day's schedule, and memories from the scientists about the trip. The Paleo- + ology link gives the reader a detailed definition along with "the scoop" on fossils. Beyond T-Rex also features Ology cards. Information about different types of dinosaurs, along with a picture, is provided on each card. Face to Fossil is a page set up like an interview. "Deena Soris" is the reporter who has an interview with a fossil of a Protoceratops. A link to the interviews provides answers to questions such as size, age, location, and how it became a fossil. Stuff to Do offers Web site visitors links to pages with dinosaur stationary, tips on how to draw dinosaurs, how to find buried bones, paleontology books, how to find fossils, and how to create your own Mesozoic museum at home. Fighting Dinosaurs offers more information about dinosaurs from Mongolia. This page has three links: See, Find and Play. See allows the reader to see what a fossil looks like through a paleontologist's eyes. The pictures are great! As you scroll over the picture with the mouse, a description of the body part is given. The body parts are even labeled. If you click on Find, you will learn possible explanations on how the dinosaurs died. Ology cards are also featured on this page. Play features a matching game of dinosaur fossils. An online quiz is featured on What You Know, featuring 10 multiple-choice questions about dinosaurs and fossils. There is even

a link that allows you to check your answers. Overall, this is an excellent interactive site for students in elementary and middle school. I would recommend this site to teachers to use as an introduction to dinosaurs.

Title: Learning from the Fossil Record
URL: http://www.ucmp.berkeley.edu/fosrec
Grade Level: K–12 and teachers
Best Search Engine: www.aol.com
Key Search Word: paleontology
Review: This is a great site to learn about paleontology. It is broken down
    into four sections: Paleontology and Science Literacy (talks about
    why fossils are important), Educational Resources, Classroom Activities, and Geological Time Scale. Each section has different articles on
    different subjects for that particular topic. There are links to definitions of any vocabulary words dealing with paleontology in the articles. There are great projects and ideas for teachers in the classroom
    activities section. There are also great projects students can do on
    their own at home or at school to enhance their understanding of paleontology. There are also many links to other great sites on other
    topics in this area. This is a wonderful site for all ages.

Title: Young Skeptics
URL: http://www.csicop.org/youngskeptics/education/lessons
Grade Level: Level 1, 5–8; level 2, 9–12.
Best Search Engine: www.yahoo.com
Key Search Words: paleontology teacher plans
Review: An outstanding site broken down into two grade levels, 5–8 and
    9–12. This animated site offers a teacher overview, length of plans
    and class periods, technical needs, navigation tools, and a student
    version of the plans. This is a progressive site with even flow from
    level to level. It is linked to the University of California, Berkeley,
    Museum of Paleontology. A check of the home page reveals numerous contact people with e-mail and phone numbers to answer questions about lesson plans.

Title: Paleontology Links, Paleontology search, Paleontology News
URL: http://quickfound.net/scitech/palentology
Grade Level: Grade 8 and up
Best Search Engine: www.dogpile.com

Key Search Word: paleontology

Review: This is an in-depth site containing information on the subject of paleontology. This site has a search section for those who wish to review additional information on this subject. This site also contains a subsection that allows for research on related subjects. The page is loaded with information on dinosaurs. The site also has a photo gallery. This site has nice pages that discuss ocean life dealing with this subject and a journal for further review. Also included in this site is paleontology news. I feel this site is good because it is full of information ranging through all the areas of paleontology.

Phenomenon

Title: Earth Science Studio and Multimedia
URL: http://www.earth.nasa.gov/Introduction/studio.html
Grade Level: Field experts and specialists
Best Search Engine: www.yahoo.com
Key Search Words: earth science and phenomenon
Review: There is a welcome site, with a brief message from an associate administrator. The next page, "who we are," describes the divisions and the people who operate them. You may click on these names or divisions. Another page explains what Earth Science Enterprise is. If you click on the link to the Earth Science strategic plan, it takes you to a page called Destination Earth: http://www.earth.nasa.gov. That link has several other links to access. There is a link for Kids Only, and Earth Science. The initial link is tied into a site map that enables you to skim down a list of sites by category. It lists sites under technology, science, practical benefits and applications, and for kids only. There is also a section for missions. You can check out satellites and classifications, such as exploratory measurements, and NOAA satellites. There is also an acronym list on this Missions page. Very cool. Now I can figure out what some initials stand for. This page could also be used when using search words "remote sensing."

Title: Weather Phenomen
URL: http://pao.cnmoc:navy.mil/educate
Grade Level: Grades 6–8
Key Search Word: phenomenon
Review: This was a difficult one! I chose Weather Phenomen because some of the other choices related to phenomenon were a little fright-

ening! This is a great site for someone just learning about different aspects of weather. Pictures and brief descriptions are provided for the following: the shape of a raindrop, natural fertilizer (lightning), St. Elmo's Fire (lightning), Diamond Dust (dangerous ice crystals in Alaska), colors of the rainbow, double rainbows, mirages, Fata Morgana (a mirage in Italy), and fog. There is also a Students-only link that was fun, easy to access, and informative about all types of weather conditions and phenomenon. There is even a site for sending e-mail. This is a great site for students to use independently if they are doing research. It is also great for teachers to allow students to explore various weather conditions. This is one of the most interesting sites I've come across so far.

Title: Oceanography from a Space Shuttle
URL: http://daac.gsfc.nasa.gov/CAMPAIGN_DOCS/OCDST/shuttle_oceanography_web/oss_cover.html
Grade Level: High school
Best Search Engine: www.metacrawler.com
Key Search Words: phenomenon and earth science
Review: This Web site captures the phenomenon of the structures of the earth's oceans. The site was developed by The University Corporation for Atmospheric Research and The Office of Naval Research. The U.S. Navy Oceanography from the space shuttle is a pictorial survey of oceanic phenomenon visible from space. The site is filled with pictures of bodies of water from around the world. Each pictorial view was taken from space and is accompanied by a description and facts. In addition to oceans, rivers, and lakes the site offers aerial views of islands, oceanic currents, water pollution, storms, and local winds. The site is very in depth and contains a variety of information. The site was easy to use and very resourceful. However, I found the majority of the information to be very confusing. The site does offer information about many different phenomena but would be most useful in a high school setting. The pictures and brief reviews are excellent examples that teachers could use during instruction.

Title: Mystery Change in Earth's Gravity Field
URL: http://www.hypography.com
Grade Level: Middle school to high school
Best Search Engine: www.yahoo.com
Key Search Words: phenomena & gravity

Review: This title was located at the Hypography and Science Tech Web site. This article describes the changes in the Earth's gravity field pre-1997 and post-1997. Scientist believe this phenomena is due to the movement of mass amounts of ice or water from the polar areas to the equatorial areas. This is a very interesting article of a timely subject since a change in the bulge of gravity fields from outward at the poles to outward at the equator has been noted by scientists since 1997. This site includes two visuals that show the gravity fields at they were, and the new shape they are moving toward. The article describes possible explanations for this change, explains why scientists have ruled some out, and why they are now focusing the reason on the oceans. This is an easy to read site for middle and high school students. It is not overwhelming with information or too technical in explanations. This article was written by NASA/Goddard SFC. Links to related information are provided.

Title: Auroras: Painting in the Sky
URL: http://www.exploratiorium.edu/learning_studio/auroras
Grade Level: Elementary school and researchers
Search Engines: www.msn.com; www.yahoo.com
Key Search Words: aurora borealis, phenomenon
Review: Developed by a teacher, Mish Denlinger, in cooperation with Jim Spadaccini and Linda Shore of The Exploration Museum, this multisensory Web site is a self-guided tour of lessons about the aurora phenomenon. The site is reliable as it is backed by SEGway, the Science Education Gateway, a national consortium of scientists, museums, and educators, and NASA Space Center. Many references are made to either NASA or the Goddard Space Flight Center scientists and their research and/or space experience, which is presented in audio clips, video clips, or movie clips. The self-guided lessons cover common questions about auroras and what they look like, what makes them happen, what the solar connection is, where you can see them, what they look like from space, and why they are different colors. There are also links to collections of awesome feature photographs of auroras on Earth and Jupiter. This Web site requests feedback, and visitors can view other users' remarks. Addresses for contact and e-mail are also available. None of the Web links was broken or inoperable. One needs Realplayer, QuickTime, and Windows Media Player in order to access the multimedia. It is worth it. One video clip took about 20 minutes to download. This site is sponsored by the regents of the University of California.

Title: StarChild
URL: http://starchild.gsfc.nasa.gov/docs/StarChild/StarChild.html
Grade Level: Grades 4–8
Best Search Engine: www.google.com
Key Search Words: phenomena and science
Review: This multimedia site has won numerous awards since making its
appearance on the World Wide Web in 1996. Woven into this site
on space and space exploration are music and videos to keep students'
interest alive while studying about the planets and the natural occurrences in our solar system. Within this Web site are pages covering
each of the planets, the sun, the moon, galaxies, stars, and other neat
space-related topics that kids love learning about. There is a special
section for educators wishing to incorporate this site into their teaching plans. Special activities accompany each title. While the layout
of this site is not as modern as some of the new science sites, it is easy
to navigate and the information is simplified so that an elementary
student can work through this site independently. The music and
videos are a nice feature as is the link to chart the phases of the moon.
Each Web page in this site has a question about the topic of the page.
There are links that lead to more information, and, finally, one that is
labeled "the answer." Students will enjoy using this Web site to learn
about the phenomena of outer space.

Title: U.S. Department of Agriculture Forest Service
URL: http://www.fs.fed.us/r4/rsgis_fire/fire_weather.html
Grade Level: Grades 9–12
Best Search Engine: www.google.com
Key Search Words: weather phenomenon
Review: I chose this site because it is something I never would have expected on a weather search. It is the U.S. Forest Service site and it is
about how weather patterns affect forest fires. I never would have
thought the weather had so much to do with forest fires. I guess that
is because I live where I do. Perhaps students in this part of the country would feel the same way I do. The opening page, the weather pattern alert page, gives a brief and very technical description of
dangerous weather patterns. Students will definitely need to learn
some serious vocabulary words before surfing this site. The introduction to this site displays some great maps and the way to understand
and navigate them. Most of the site contains maps, archival information and maps, weather related links, and news. This site is definitely
for the secondary student and could be fun and interesting.

Title: Phenomenon
URL: http://www.stateoftheheart.n/phenomenon/
Grade Level: Grade 8 and up
Best Search Engine: www.ixquick.com
Key Search Word: phenomenon
Review: This Web site is designed for children in grade 8 and up to explore. This Web site offers projects, products, and much more. The site's main page has "Enter here." Users will need a Flash player, which can be downloaded at no cost. On every page are "moving icons to explore." A click of the right mouse button allows the user to explore any of the pages: Links, Books, Extraterrestrial, Prophecies, and much more! I explored Cults. I found that there were many cults to explore individually. I found this Web site to be a little advanced for a child below 13 years old. It is suitable for children and professionals for gaining a basic understanding of phenomena and how they affect us. Furthermore, this Web site overall is beneficial for all students and educators, and can be used with caution and parental guidance. The site is updated frequently and encloses a privacy statement and no disclaimer for validity.

Title: The Phenomenon of Science, a Book on MSTT
URL: http://pespmc1.vub.ac.be/POSBOOK.html
Grade Level: High school and college
Best Search Engine: www.yahoo.com
Key Search Word: phenomenon
Review: The *Phenomenon of Science* is a book that deals with a cybernetic approach to human evolution. It begins with Metasystem Transitions Theory and ends 14 chapters later with Integration and Freedom. Throughout the 14 chapters it deals with a slew of topics revolving around science and humans. The book itself is no longer in print but you can pull up the entire book from this site.

Title: Beyond Discovery: The Path from Research to Human Benefit
URL: http://www.beyonddiscovery.org
Grade Level: Middle and high school
Best Search Engine: www.google.com
Key Search Word: phenomenon
Review: This Web site offers an 11-page article on the Ozone Depletion Phenomenon. It discusses "the discovery of the effect of chlorofluorocarbons in the Earth's atmosphere and the subsequent discovery of the ozone hole in the Earth's atmosphere." Also offered are links to related

Web sites, a visual satellite image of the hole in the ozone layer, a timeline tracing the discovery of the ozone, first identified in 1840, through to the complete ban on the industrial production of CFCs in 1996. Interestingly, this site is also available in Chinese! I did like the glossary of related terms. I found it to be a good reference article for a middle or high school student researching this timely topic.

Title: Asteroids
URL: http://nssdc.gsfc.nasa.gov/planetary/text/asteroids.txt
Grade Level: Middle and high school
Best Search Engine: www.askjeeves.com
Key Search Word: phenomenon
Review: The article gives the reader a substantial amount of information about asteroids. The article explains that asteroids are "Metallic, rocky bodies without atmospheres that orbit the sun, but they are too small to be classified as planets." The article also mentions that the largest asteroid is called Ceres. There are three types of categories that asteroids fall into: C-type, S-type, and M-type. The reader can find out what type an asteroid is, its color, and its albedo (reflective power). The reader can also find out if the asteroid inhabits the inner, outer, or middle region of the asteroid belt. A fact sheet lists the names, diameter, mass, rotation period, orbital period, spectral class, and the semimajor class of the asteroids. One would learn a lot of useful information if they had to do a paper on this particular topic.

Plate Tectonics

Title: Classroom of the Future
URL: http://www.cotf.edu/ete/modules/msese/earthsysflr/plates1.html
Grade Level: All grade levels
Best Search Engine: www.yahooligans.com
Key Search Words: plate tectonics
Review: This is a very well organized and easy-to-use site. There are many resources for students and teachers alike. This site is not geared toward one or the other; it is right down the middle. There are lots of fun activities for the children to do while learning. I could not get in to the Teacher's Lounge because I did not have a password. Take a look at this site.

Title: Geology: Plate Tectonics
URL: http://www.ucmp.berkley.edu/geology/tectonics.html

Grade Level: Grade 6 and up
Best Search Engine: www.metacrawler.com
Key Search Words: plate tectonics animation
Review: Good animation of the continental drifts over the history of the earth. You can see how much the continents have moved over the years. Also with each animation are links to descriptions of each of the four major paleontological eras. And within each page are even more links to relevant terms and definitions that go with each era. Very informative, but not so much information that you feel overwhelmed. Also very good job of putting into perspective the amount of time over which our planet has evolved.

Title: Geology: Plate Tectonics
URL: http://www.ucmp.berkeley.edu/geology/tectonics.html
Grade Level: High school and above
Best Search Engine: www.earthlink.com
Key Search Words: plate tectonics
Review: As you enter the site you will see a picture of the plate tectonics. There is a brief description of the new understanding of the Earth and the different theories. Under this section you can click on the link to view the history of plate tectonics or learn about the mechanisms driving plate tectonics. The history of plate tectonics discusses the rock history of an idea. The page begins with Alfred Wegener's theory and discusses the continents up to now. If you click onto Alfred Wegener's name you will get a biography about him. As you finish reading the section the site gives you a list of books to learn about plate tectonics. The mechanism link discusses the main features of plate tectonics, mid-oceanic ridges, geomagnetic anomalies, island arcs, and deep sea trenches. This Web site allows you to see plate tectonics animation. If you click onto any one of the animation sites it will bring a picture of the world and what has happened through the time. Each link gives you information on Cenozoic, Mesozoic, Paleozoic, and Precambrian times. This site gives you a variety of ways to learn about plate tectonics. While visiting this site you will get an understanding of what has happened from way back when.

Title: Plate Tectonics
URL: http://www.ajkids.com
Grade Level: 8–12
Best Search Engine: www.metacrawler.com
Key Search Words: plate tectonics

Review: This page is a part of the Ask Jeeves Web site. Students can type in the search words plate tectonics to receive great information on this topic. The page is divided into four sections: What is Plate Tectonics?, Evidence of the Existence of Plate Tectonics, Types of Boundaries, and The Growth of Continents and Plate Tectonics. The site presents information about plate tectonics in a clear and easy-to-read manner. The most comprehensive part of this site is the section on boundaries. This section has myriad useful pieces of information and pictures. Key vocabulary words are linked so that the students can get background information and definitions. The page has some pictures, mostly of maps, but more pictures would be helpful to the students. This site offers links to pages about volcanoes, earthquakes, and rocks and minerals. This site is good for an overview of plate tectonics. However, students looking for in-depth information on plate tectonics should not solely depend on this site.

Title: All about Plate Tectonics: Earth's Plates and Continental Drift
URL: http://www.enchantedlearning.com/subjects/astronomy/planets/earth/Continents.shtml
Grade Level: Elementary and middle school
Best Search Engine: www.yahooligans.com
Key Search Words: plate tectonics
Review: This site is part of the Enchanted Learning Web site. It is a great site for elementary and middle school students to learn the basics about plate tectonics. The site begins with a brief description of how the earth was formed. Key words and phrases are linked to definitions throughout the section. If you click on the Earth's mantle, you are taken to another page that includes a printable diagram of the interior of the Earth that can be labeled. The page also includes a lot of information about the inside of the Earth, along with a lot of colorful, labeled diagrams of the layers, inner and outer, and the crust. At the bottom of this page are links to related sites, such as Earth diagrams, soil layers, and volcano diagrams, all of which can be printed and labeled. Back on the main page, there is a great animated picture of the continental drift. There is also a great image of the Earth's major plates. The pictures are labeled and easy to read. Plate tectonics is explained farther down on the page. Pictures and diagrams are also featured. Animated pictures help illustrate divergent and convergent plate movement as well as lateral slipping plate movement. At the bottom of the main page, there is a list of related activities, including an interactive quiz, as well as several links about the continental plates. I think this

site is a great introduction to plate tectonics for elementary and middle school students. The language is kid friendly and the pictures are great, especially the animated ones. I think teachers will like this site also for its printable diagrams, which can be used during a lesson.

Title: USGS
URL: http://www.geology.er.usgs.gov
Grade Level: Middle school and high school
Best Search Engine: www.metacrawler.com
Key Search Words: plate tectonics
Review: This is a good Web site. It has graphics, and descriptions of the planet before, during, and ideas of how the planet will move in the next few centuries. The links are updated but not frequently. The government takes care of the site and lists an address of where to be reached if there are any questions.

Title: The Dynamic Earth, The Story of Plate Tectonics
URL: http://pubs.usgs.gov/publications/text/dynamic.html
Grade Level: Grades 8–10
Best Search Engine: www.yahoo.com
Key Search Words: teacher lesson plate tectonics
Review: This Web site is geared toward grades 8–10. It is broken into eight sections from history to plate tectonics and people. The site is well organized and flows well. Overall, it is a good site with clear and concise explanations as well as diagrams and graphics. The complete text of this article is available for purchase off the Web site. The U.S. Geological Survey updates the site and can be reached through contacts listed on the site.

Title: The Wilson Cycle and a Plate Tectonic Rock Cycle
URL: http://csmres.jmu.edu/geolab/Fichter/Wilson/Wilson.html
Grade Level: High school and above, with some pages being quite advanced
Best Search Engine: www.yahoo.com
Key Search Words: plate tectonics
Review: This is a great Web site with lots of high-quality illustrations that explain the movement of tectonic plates. There is a self-test to check your comprehension of the site. The different types of plate boundaries are discussed, with very nice illustrations of each. The various pages that make up the site are linked in various ways. This permits the visitor to go directly to a stage/page they wish to view or to see

the entire cycle in chronological order. This is a must-see for students and teachers of Earth Science.

Title: Life in the Universe: Plate Tectonics
URL: http://www.lifeintheuniverse.org/noflash/Platetectonics-05-02-01.html
Grade Level: High school and above
Best Search Engine: www.google.com
Key Search Words: phenomenon plate tectonics
Review: This Web site began by defining plate tectonics. Then it gave information about sliding plates, tectonic activity at plate boundaries, and the important role of plate tectonics, stating that without plate tectonics there would be no life. While this site gave a great deal of information, glossary terms, and links to the global carbon cycle, the greenhouse effect, and Mars's history, there were few photos, no graphs, diagrams, or charts, although this may have been because the computer used to access this site did not have Flash. I think this Web site would be helpful if you were looking to get a quick overview of plate tectonics and associated terms. However, I do not think it would be enough if it were your only site. It is a good starting point.

Title: The ABC's of Plate Tectonics—Introduction
URL: http://webspinners.com/dblanc/tectonic/ptABCs.shtml
Grade Level: High school and above
Best Search Engine: www.metacrawler.com
Key Search Words: plate tectonics
Review: The opening page of this Web site has a table of contents with six headings: Preface: Introduction to the ABC's of Plate Tectonics; Remedial Reading: The Basics of Plate Tectonics; Lesson # 1: Buoyancy and Floating Continents; Lesson # 2: Sedimentation and Continental Growth; Lesson # 3: When Continents Collide; and Lesson # 4: The Mechanism of Plate Tectonics. The Introduction explains what this Web site presents: a broad analysis of the basic principles that should apply the movement of plates, some new hypotheses about how they apply to convection and landform formation, and some expected scenarios for differing tectonic events. At the end of the introduction there is a link to the U.S. Geological Survey Web site, and you are instructed to go there for a more comprehensive explanation of plate tectonics. You can click on the link to the next topic at the bottom of this page or you can return to the top and find the topic of your choice. All of the links lead to more in-depth discussion of the topic. This Web site

is authored and maintained by Donald L. Blanchard. He does not explain who he is, but if you click on the link at the very top of the home page you are taken to his personal Web site with articles he has written. As I was exploring this site I thought the information in it was good, but at the end of the last lesson is a note that says, "The opinions expressed in these lessons are those of the author . . . and probably (in fact quite likely) do not coincide with those of any professional authority on plate tectonics or paleogeography." That said, I'm not sure I would recommend it to anyone without a background in this science as most people wouldn't be able to separate fact from his hypotheses. Someone with a good background in the field might find it a valuable source of alternative views.

Title: Plate Tectonics: The Worst Natural Disasters
URL: http://www.thinkquest.org/J002319
Grade Level: Grades 4–12
Best Search Engine: www.aol.com
Key Search Words: plate tectonics
Review: This is an in-depth site on plate tectonics. There is an overview of plate tectonics, then there are more in-depth sections on earthquakes, tsunamis, and volcanoes. Each section talks about the causes of earthquakes, tsunamis, and volcanoes, then there are quizzes on each topic. In the discussion of each topic, there are links to definitions to vocabulary words dealing with plate tectonics. There are also experiments that go along with each topic. These experiments can be done at home or in the classroom. There are great pictures to show the disasters caused by earthquakes, tsunamis, and volcanoes. This is a great resource for those just beginning their studies on plate tectonics and also a great resource for teachers as well.

Remote Sensing

Title: ASPRS: Imaging and Geospatial Information Society
URL: http://www.asprs.org
Grade Level: Interested parties
Best Search Engine: www.yahoo.com
Key Search Words: remote sensing
Review: This organization's Web site lists conference dates and locations. It has links on the left denoting membership opportunities, meetings, publications, site search news and external affairs, resources, society information, addresses, and other little goodies. There is a section for

remote sensing core curriculum. There are annual indexes listed on the right side of the page, as well as the current issue of the magazine, *PE&RS*. The core curriculum has five volumes in it: 1. Introduction to Photo Interpretation and Photogrammetry, 2. Overview of Remote Sensing of the Environment, 3. Introductory Digital Image Processing, 4. Applications in Remote Sensing, and 5. K–12 Education. I went into volume 3. It says it is under construction, but I went into weather and it took me to a whole slew of sites. You can get commercial weather sites, federal weather, and educational weather sites. I am going to check out CNN and see when this weather is going to clear up. Sunny in two days.

Title: The Remote Sensing Tutorial
URL: http://rst.gsfc.nasa.gov/start.html
Grade Level: Grade 7 and up
Best Search Engine: www.ixquick.com
Key Search Words: remote sensing
Review: This is another NASA science site. It also contains numerous links to other remote sensing sites on the Web. The tutorial itself contains 21 sections on various types of remote sensing from image processing to cosmology and remote sensing on Earth and in outer space. This Web site on remote sensing examines the inward and outward look at the Earth, planets, and galaxies. "This Overview serves two main purposes: To provide a brief synopsis of the uses and history of remote sensing and to describe the contents of the entire Tutorial with suggestions on best ways to utilize it (primarily when accessed as a Web Site)." It ends with a quick look at the latest products now being acquired by commercial remote sensing satellites. We strongly recommend that you read through the entire overview as a proper introduction to both the tutorial and to the many fields of practical activities involving the principal ways in which remote sensing and allied fields contribute to gathering information about the surfaces and atmospheres of the Earth and other solar planets. Taken directly from the Web site: "The Tutorial has been developed for prime users such as faculty and students at the college level; science teachers at the High School level; gifted or interested students mainly from the 8–12 grade levels; professionals in many fields where remote sensing comes into play. It is also developed for individuals who need insights into what this technology can do for them; that segment of the educated general public who is curious about or intrigued with the many

accomplishments of the space program that have utilized remote sensing from satellites, space stations, and interplanetary probes to monitor and understand surface features and processes on Earth and other bodies in the solar system and beyond." The principal author of the tutorial is Dr. Nicholas M. Short. There are several points to consider when looking at this tutorial. First, the summary and preview button. found on each page, is bounded by blue lines. Secondly, it is prepared for online display and needs a balance between text and illustrations with a monitor screen setting of 600 by 800 pixels. It is suggested that the tutorial is large and needs to run on a 56k or faster modem. Lastly, there are many links to the tutorial and for the most part the image is on the last page or starting page for accessibility. Other sections of the tutorial cover the history of remote sensing and include an appendix. While this is not a glitzy site, the information presented is straightforward and easy to understand. Various charts and graphs are also used to further explain topics. It is easy to navigate, and all links were working at the time this site was accessed.

Title: Ocean Remote Sensing
URL: http://fermi.jhuapl.edu/
Grade Level: High school
Search Engine:www.yahoo.com
Key Search Words: earth science & remote sensing
Review: This site is sponsored by the Johns Hopkins University Applied Physics Laboratory and would be of use to high-level high school students who are interested in images and information on current research in the area of remote sensing related to oceans. This site provides numerous remote sensing images collected through research projects. Some of the information includes Atlantic Hurricane track maps and radar images beginning in 1995 through the current hurricane season. NOAA sponsored the hurricane images in addition to images of wind fields and wind speeds in Alaska and the ocean off of the north Atlantic states (New Hampshire to North Carolina). This site also provides information about current projects conducted by Johns Hopkins and funded by NASA. These projects provide visual images and text covering how research is being conducted and preliminary information on results. One of these research projects is a test of a D2P radar remote sensing device over the ice sheets of Greenland. This is an interesting site to explore and can provide specific information some students may be searching for when working on a project. Some of the information

available was current, but the site also included historical information going back many years. Some of the images are blocked with password access only and some were not working.

Title: Md-Lasertech
URL: www.md-lasertech.com
Grade Level: Adult
Best Search Engines: www.metacrawler.com; www.google.com
Key Search Words: remote sensing
Review: This site is actually an advertisement for Md-Lasertech, a provider of high-quality remote sensing services and technology. It gives a general profile of the company and its products for sale and service. One of its pages is called About Remote Sensing. For those of us who had no idea what remote sensing is, it explains it very well. It also give a history of RS, the theory of operation and technological advantages. It includes a figure on radiation power and emission wavelength. Some of the information is technical and some is elementary. The technical sections are well explained and cover long-path photometry, pollutants of vehicles (emissions), and light beams indicating concentrations of gases such as carbon monoxide, hydrocarbons, and nitrogen oxides. Remote sensing is also used to measure speed and acceleration and can record images of license plates and the rear of a vehicle for law enforcement. The site has wonderful displays of equipment used for speed limit control. Tunable diode lasers (TDLs), which are small crystals, play a role in the development of RS. Lockheed research, GM cars, and certain states' contracts for emission control and speed limit control prompted the growth of remote sensing. A very interesting article.

Title: Canada Centre for Remote Sensing
URL: http://www.ccrs.nrcan.gc.ca/ccrs/homepg.pl.?e
Grade Level: High school
Best Search Engine: www.yahoo.com
Key Search Words: remote sensing
Review: I had to go north to find this site. Home page links are to several sections: About Us, Research and Development, Imagery and Sensors, Learning Resources, and Community. It also offers information on Climate Change, Northern Development, Hyperspectral Techniques, and Natural Hazards. The Web site also offers a What's New area where the user can learn about the new 5. scale model and projects at the innovation acceleration center. This site is a wealth of information.

Title: Geoscience and Remote Sensing Society
URL: http://www.ewh.ieee.org/soc/grss/
Grade Level: High school and college
Best Search Engine: www.dogpile.com
Key Search Words: remote sensing
Review: This Web site, created by Webmaster Dr. Bill Emery of the University of Colorado and designed and maintained by Tara Jensen, "seeks to advance geoscience and remote sensing and technology through scientific, technical, and educational activities." It was created to promote the exchange of information through meetings, workshops, publications, conferences, and so forth. It serves as a good research tool as it offers a monthly journal called *Transactions on Geoscience and Remote Sensing* (TGARS). This site also contains the newsletter of the Remote Sensing Society. It is a sound resource for current articles in this field for the advanced learner.

Rivers

Title: Save Your River
URL: http://www.education-world.com/a_lesson.html
Grade Level: K–3
Best Search Engine: www.yahoo.com
Key Search Words: remote sensing
Review: This was a good Web site of approximately four pages. This site offered brief definitions of rivers with objectives and materials for one to three lessons, depending on time availability. Class discussion was offered in small groups, with individual projects and posters to follow. What was nice was you could enter your state and find rivers where you could teach to make the lesson relevant. There were no links to contact the author, and there was a lesson plan Web site search.

Title: Rivers
URL: http://www.nps.gov/rivers
Grade Level: College and above
Best Search Engine: www.google.com
Key Search Word: rivers
Review: This site connects you to the National Wild and Scenic River system. As you navigate through the site there are 11 links: Designated WSR, River information, River and trails, WSR Council Publication, Study rivers, About WSRs, Guidelines, Site index, Agencies, WSR Act, and NRI. The Designated WSR has a list of 40 states with rivers.

If you click onto a state the site will provide information on the designated reach and mileage along with a related site. The river information gives a list of organizations to contact about rivers and river management. The river and trails link talks about the program and how they conserve rivers, preserve open space and develop trails. This link is updated every month. The WSR Council (National Wild and Scenic Rivers) describes the NWSR act and information on which rivers are included and the four agencies that handle these rivers. The publication link provides information on rivers that you can download. The Study River links gives information on the public law 542 and the eight rivers that were in the initial components as well as the 148 rivers added since then. About WSR discussed when the country realized that rivers were being damaged and how Congress created the National Wild and Scenic River System. The guideline gives you a federal registrar paper from September 7, 1982. The site index lists bibliography, environmental quotations, questions, rivers and water facts, lengths, agencies and a variety of other links. The agency links go to the four federal state agencies that have Web sites to provide information. The NRI link gives you information on the 3,400 free-flowing river segments in the United States. This site is quite extensive and provides a lot of knowledge about rivers.

Title: American Rivers
URL: http://www.amrivers.org
Grade Level: All grade levels
Best Search Engine: www.metacrawler.com
Key Search Word: rivers education
Review: This site was one of the best I've come across so far. The intent of the site is to educate people about rivers, their purpose, and their importance to the environment. The tutorials on rivers are simple and complete broken down into easy-to-follow parts. This site contains definitions, games, and experiments, all usable in the classroom. This site also includes issues facing rivers: dangers, campaigns, communities, and action groups working to protect our rivers. Whether you are just teaching grade school kids about rivers or organizing your own campaign to protect rivers, this site has it all.

Title: Athena Review
URL: http://www.athenapub.com/rivers1.htm
Grade Level: All grade levels
Best Search Engine: www.yahooligans.com

Key Search Words: rivers

Review: This Web site does not tell much about rivers, but it has some very interesting and useful background information and photographs that are useful. I would recommend this site for anyone looking for visuals for a project or just want to see some cool pictures of rivers as viewed from space. Each picture is coupled with a brief history of the river. Check this out! It is really cool.

Title: All Along a River
URL: library.thinkquest.org/28022
Grade Level: Grade 11 and up
Best Search Engine: www.metacrawler.com
Key Search Words: rivers and kids

Review: This site was developed by ThinkQuest Inc., which is a non-profit organization that offers programs designed to advance education through the use of technology. This is a wonderful interactive site that students and teachers can use to learn about rivers. The site is divided into multiple sections. The first three sections deal with river erosion, river volume, and river velocity. These sections offer great information and give students excellent graphics through the use of diagrams and pictures. Another section allows students to see rivers of the world. The students click on this link and they are taken to a map of the world. The names of world rivers can be easily located on the map. When the student clicks on a river, they are taken to a page that has information on that particular river. The most interesting section on this site is Along the River. The students can click on different parts of the river to learn about such things as rapids, waterfalls, and river canyons. This site also offers a great tool for teachers. The Online Study Area is a place where teachers can go to learn how to carry out lessons online. It also allows for students to do homework online. This site is informative and has great graphics.

Title: Geography Action! Rivers 2001
URL: http://www.nationalgeographic.com/geographyaction/rivers/
Grade Level: Elementary and middle school
Best Search Engine: www.yahooligans.com
Key Search Word: rivers

Review: This is a really good site. The main page features an interactive river system. When clicked, you are taken on a journey through a river system. The map features labeled diagrams of a river. When you click on the different parts of the river, a definition pops up. Some of the terms are tributaries, river source, main river, upstream, downstream, river mouth, meanders, floodplain, and watershed boundary. At the bottom of the map, there is a section for teachers that offers lesson ideas, along with related handouts. Adobe Acrobat Reader is required in order to download the handouts. The handouts include a river system illustration that can be labeled, and a list of river system words and definitions. Take Action is another link on the main page that gives readers plenty of activities related to river conservation. There is also a calendar of upcoming events in your area. The calendar includes detailed event information and contact numbers. The Activities Menu gives several ideas for teachers and students. There are experiments, creative writing, and online games. Other activities include building a watershed, making a collage, or building a wetland. Each activity includes easy-to-follow, detailed instructions. I really liked this site. It is interactive and provides a lot of resources and activities for teachers to use to enhance a unit about rivers.

Title: River Resource—lists of links, bibliography, databases of classes studying rivers
URL: http://www.riversource.com/default.html
Grade Level: K–12
Best Search Engine: www.metacrawler.com
Key Search Word: rivers
Review: River Resource bills itself as "a gateway for productive internet exploration." It is intended to serve as a connection to "facts, books, and people studying rivers." The home page has four links: Philosophy Behind These Pages, Links to River Pages, Bibliography of River Books, and Connecting with Other River Classrooms. In addition, on the bottom of the home page is a Search For box that allows you to type in your own topic. The Philosophy page gives links to other Web pages with information on rivers, a bibliography of books, and classes who are studying rivers. In addition, there is a link to the Web site of the author. Clicking on his name leads to information about his background and information about rivers he has explored. He also has information about how he writes his books. There is a link to writing assignments aimed at helping teachers and students learn how to write about rivers. This page contains wonderful links for teachers. Clicking on the Links to Rivers pages takes you to a page where you can go to river specific links or links of general river information. There is an additional link to an alphabetic index that will link you to specific rivers. The general river information page link is

divided into topics. When you go to that page there is a wealth of topics to choose from ranging from river curriculums to Music, Art, and Film. The Bibliography of River Books contains lists of books related to rivers. Some of them give specific ages. The author states that this site is under construction. There is a link for adding books to his list. The Connecting to Other Classrooms leads to a site where teachers can communicate with other teachers about class river projects or find other classes studying rivers. On the link to finding other classes there were only five projects posted. However, if a class was doing a river project this would be a great way to share it. Overall this is a great site for teachers and students.

Title: Water Watch Current Water Resources Conditions
URL: http://water.usgs.gov/waterwatch
Grade Level: Grade 6 and up
Best Search Engine: www.yahoo.com
Key Search Word: rivers
Review: This site, maintained by the U.S. Geological Survey, has some nice maps relating real-time stream flow data expressed as a percentage of average for that specific day. Students can also use drop-down menus to view maps of the states and check them for drought or flood conditions. The site does a great job of explaining the data, and I feel that a science teacher could come up with some excellent lessons using the varied data available here. However, as a standalone experience this Web site does not offer much value.

Rocks

Title: Discover How Rocks Are Formed
URL: http://www.fi.edu/fellows/payton/rocks/create
Grade Level: Teachers
Best Search Engine: www.yahoo.com
Key Search Words: rocks
Review: From the get-go, you can see how rocks are formed. It gives you the option to click on sedimentary, metamorphic, or igneous rocks. Each link offers a simplistic explanation for the forming of rock. It also gives the viewer a chance to look at a photo of a sample of rock formed in that category. If you click on the hard hat, it takes you to the Rock Hound Home Page. This has lesson plans, literature, activities, and other things under the Teaching Connections section, on the right of the screen. The items on the main section of the page ap-

pear to be for kids. Maybe not kids, but it is for at least Earth Science students. I looked at a few questions, and decided I didn't need another award! There is a rock quiz, and a pebbling puzzle, and a few other sites here. The Puzzles section has crossword puzzles, word searches, and a jigsaw puzzle. (Which I found frustrating. LOL.) You may find the teaching connections more valuable, because within the lesson plan, the author included additional Web sites to view in preparation of lessons.

Title: Rocks and Minerals
URL: http://edtech.kennesaw.edu/web/rocks.html
Grade Level: Grades 7–8
Best Search Engine: www.metacrawler.com
Key Search Word: rocks
Review: This Web site is an excellent resource for both students and teachers. The site recently won an award for Internet Scout Project. The site has information on various types of rocks, minerals and caves. The site has numerous links that lead students and teachers to information on rocks and minerals. The site provides explanations and photographs of a number of rocks and rock formations around the world. Teachers can also access this site to find lesson plans on all areas of rocks. There are activities for parents, teachers, and students on this page. The page is organized into sections, which offer information on web pages, lesson plans, and activities. The page is well maintained and easy to access. The Web site can be used for a variety of grade levels.

Title: Rocks for Kids
URL: http://www.rocksforkids.com
Grade Level: Grade 5 and up
Best Search Engine: www.ixquick.com
Key Search Word: rocks
Review: This site seems to be user friendly and easy to read. The Rocks for Kids page has a "Rocks" table of contents page. I chose to explore only this page. I found that entering with a click of the mouse led me to valuable information about rocks. The subtitles (arranged in chapters 1–15) were how rocks are formed, the Earth's crust, rocks, minerals, crystals, and soil, and rock cycle, ingenuous, sedimentary, and metamorphic rocks. Thus, I explored how rocks and minerals are formed. This subsite listed valuable and recent information in a table format. I noticed a subsite called "The stupid page of rocks." I discov-

ered that there were three "rocks" to open and obtain more information. I found this Web site to be valuable for children who are working on a report and need information about rocks and their formations. Teachers and other educators would find this site to be helpful and useful for any science curricula. Furthermore, this site seemed appropriate for grades 5 and up.

Title: Antarctic Geology: The Rocks Beneath
URL: http://www.aad.gov.au/information/more_res/geology.asp
Grade Level: Elementary and above
Best Search Engines: www.metacrawler.com; www.msn.com
Key Search Words: rocks, rocks/Antarctica/Antarctica geology
Review: The purpose of this Web site is quoted as "Antarctica, valued, protected, understood." It provides scientific expedition and research information on Antarctica including prehistory and mining. It is written to inform interested parties that the rocks of the East Antarctica Shield are among the oldest known rocks on earth, as old as 4 billion years; that millions of years ago Australia, Antarctica, and South America were one land mass, Gondwana Land, and that fossil discoveries imply that Antarctica was not entirely covered by ice as recent as 3 million years ago. Antarctica is larger than Europe and nearly twice the size of Australia, at 13 million square kilometers. Antarctica yields evidence of having had dinosaurs, amphibians, marsupials, and similar plant spices of Australia, fossilized spores and pollen of sundews, Tasmanian myrtle, deciduous beech, fern, and pine. The Nunataks, or Transantarctic Mountains, are among the world's longest and most impressive mountain chains. There are several seas in Antarctica, the Ross Sea to the northwest and the Weddell Sea to the southwestern corner. Antarctica has a peninsula and there has been major volcanic activity. Mining seems to be a very touchy subject in Antarctic. There are minerals of coal and iron ore. Current difficulties associated with the recovery of mineral deposits could be overcome with time and technology and therefore in order to prevent mining exploitation, Nations of the Antarctic Treaty System agreed in 1991 to put a halt to the exploitation of minerals, which banned mining indefinitely and was enforced by 1998. The authors of this article, Wendy Rockliffe and Patrick Quilty, are in some way involved with the Australian Antarctic Division but no further information about them is given. There are numerous contact vehicles, mailing addresses, e-mails to certain divisions, and phone numbers. The Australian government published this article and the domain is government. The AAD is headquartered at the Aus-

tralian National Antarctic Research Expeditions in Kingston, Tasmania. The home page of the overall AAD site has links that are updated daily and weekly, including a link to This Week in Antarctica. All of the links were accessed easily and no additional software was required. No author opinions were expressed in this rock review. Only scientific information presented in a well-organized and easily understood manner. Some graphics were used and videos and virtual reality links were included throughout the expedition links. This part of the world is fascinating.

Title: Robotic Antarctic Meteorite Search
URL: http://www.frc.ri.cmu.edu/projects/meteorobot2000
Grade Level: Middle and high school
Best Search Engine: www.yahoo.com
Key Search Words: rocks & earth science
Review: This is a very interesting Web site that is sponsored by NASA, the Robotics Institute, and Carnegie Mellon University. This site provides information about research being conducted in the Antarctic using the Nomad, which is a robot that will search for meteorites and automatically classify the rock samples. The Nomad has been able to classify indigenous meteorites and terrestrial rocks. This site describes the Nomad and its hardware, such as the rock classifier. The expeditions Nomad has already completed are described, including how they were conducted and results. This site provides links to sources of related information and sponsors. Another feature is a link to abstracts of journal articles that have been published about this research project. This site is easy to navigate and was updated with research information from recent expeditions.

Title: The Essential Guide to Rocks
URL: http://www.bbc.co.uk/education/rocks
Grade Level: Grades 3–5
Best Search Engine: www.google.com
Key Search Word: rocks
Review: This Web site was full of information. There were colored diagrams, maps, and photos. It was divided into the following categories: Stones at home, Walks with rocks, Program archive, Links/resources, and the Presenters. I found Stones at home to be most interesting. There were rock cycle experiments, directions on how to make a sedimentary sandwich, geology of the bathroom, the graveyard rockwall, and how to make rock art. This is definitely a site I think students

would enjoy checking out. It had many hands-on activities and practical examples.

Title: Discover How Rocks Are Formed
URL: http://sln.fi.edu/fellows/payton/rocks/create/index.html
Grade Level: Elementary school
Best Search Engine: www.yahooligans.com
Key Search Word: rocks
Review: This site has three links on its main page. There is a great picture that shows the different kinds of rock beneath the Earth's surface. By clicking on the different areas of the picture, you can watch these animations that will show you how the three kinds of rocks are formed. At the bottom of the page there are links to sedimentary, metamorphic, and igneous rocks. When you click on sedimentary, you are taken to another page that features an animated picture. There is a brief yet informative description of how these rocks are formed. There are also links at the bottom of the page that offer detailed information about the different kinds of sedimentary rocks, such as sandstone, limestone, shale, conglomerate, and gypsum. By clicking on each type of rock, you will be taken to yet another page that offers a picture of that type of rock along with a brief description. The other two links offer the same type of information related to metamorphic and igneous. Both pages offer additional links that show pictures and descriptions of the types of rock in these categories. This is a great site for elementary school students. The kids will enjoy the animated graphics. The teachers will enjoy the "kid-friendly" terminology.

Seismology

Title: Seismology Word Search
URL: http://freebies.about.com/library/words/blws013.htm
Grade Level: Middle and high school
Best Search Engine: www.metacrawler.com
Key Search Word: seismology
Review: This is not an actual Web site but it was a fun activity to reproduce for any student who is studying seismology.

Title: Seismology
URL: http://www.geo.mtu.edu/seismology/ups/waves.html
Grade Level: Grades 4–6

Best Search Engine: www.google.com
Key Search Word: seismology
Review: A good Web site with terminology broken down into specific areas. Almost any term can be built on for greater study depending on lesson timelines. Many topics were covered, including body waves, surface waves, primary waves, secondary waves, Love waves, and Rayleigh waves. There was a link to contact Michigan Technological University as well as the author.

Title: Seismology
URL: http://home.wish.net/~riknl/#A
Grade Level: Grade 9 and up
Best Search Engine: www.yahoo.com
Key Search Word: seismology
Review: This is a nice Web page with text links throughout. There are also some nice illustrations. Text does a great job explaining the illustrations, and a table provides information on the Mercalli scale. Links are provided for researching recent seismic events. There is not a lot of information on this page but everything works well and is nicely done. This is an excellent site to start with in your quest to understand seismology.

Title: Seismology Research Center
URL: http://www.seis.com.au
Grade Level: Graduate
Best Search Engine: www.google.com
Key Search Word: seismology
Review: As you click onto the site you will see the title about the SRC and two columns. The SRC title gives you the history and what kind of work the center is concerned with. Within the link it talks about equipment and analysis of software. Each description allows you to click for further information. The site also discusses the earthquake preparation, alarm and response, seismograph network operation, consulting, and the Philippines project. On the left side of the page it gives you products: digital recorder, digitizer, drum recorder, Power Point presentations, and applets. Each one of these products can be clicked on for more information and a description of the equipment. The right column has services: seismic monitoring, earthquake alarms system, hazard consulting vibration monitoring, and a link for registered users. Under each of these columns there are two subheadings. One is earthquakes and the other general. The earthquake links

you to the Introduction to earthquakes, Australian earthquake news, World earthquake news and Have you felt an earthquake? The general column allows you to contact them and gives a calendar of events, links to other organizations and Web site details. The site is quite interesting since it's based in Australia.

Title: Geology Labs On-Line
URL: http://www.sciencecourseware.com
Grade Level: High school
Best Search Engine: www.metacrawler.com
Key Search Words: seismology education
Review: This site contains a virtual earthquake program. It is an experiment that is completely online. The program models the experimental information from which the students are expected to gather their data. There are printable results for in-class demonstrations, and all information can be saved directly on the site for the students' future use, which allows a group of kids in one group to work independently, without having to worry about who has the disk or whose computer it's saved on. It seems as though it could be interesting. The site also has example experiments for student viewing as well as tutorials on the major topics of the experiment. I would venture to guess that more advanced students could be left alone to complete this experiment independent from teacher guidance.

Title: Earth and Plate Tectonics
URL: http://www.geology.about.com
Grade Level: intermediate and secondary
Best Search Engine: www.metacrawler.com
Key Search Words: seismology and education
Review: As I looked through this Web site, I found some interesting links. The site informs you about every aspect of earthquakes both in theory and practice. They also have accounts of past earthquakes as well as pictures. They also have a link where you could experience an earthquake through computer animation. I didn't, however, see any updated information or contact person.

Title: Harvard Seismology
URL: http://www.harvardseismology.edu
Grade Level: College
Best Search Engine: www.dogpile.com
Key Search Word: seismology

Review: This site is jammed with information. It is a site that I don't think children will understand, because the reader must have some prior knowledge to fully comprehend this site. The site has five major links. The first link is Contact Information/Related Sites, which takes the reader to a page that has additional names of contact people at Harvard University. The link further lists other links for additional research. Another link allows the reader to a CMT Catalog Search. The reader can search a catalog to find other information. This site also offers information on graduate level and undergraduate level courses. Further information is available from online data. The site that contains the most information is the research link, which has information on Surface Waves, Wave Propagation, Normal Modes, 3-D Earth Surface, Convergent Margins, Geomap, Earthquake Dynamics, Event Locations, Unusual Earthquakes, The Inner Core, and the Centroid Moment Tension Project. I think you will enjoy researching this site.

Title: Seismological Laboratory
URL: http://www.gps.caltech.edu/seismo/seismo.page.html
Grade Level: Researchers, K–12 teachers
Best Search Engine: www.metacrawler.com
Key Search Word: seismology
Review: This Web site is authored by the Seismological Laboratory at the California Institute of Technology. The home page contains links to eight major topics: Academics, Earthquakes, Seminars, Seismolife, People, Research, Preprints, and Late Breaking. The Academics page lists the professors and courses available at Cal Tech. Useful information if you plan to go there; otherwise it can be skipped. Earthquakes links you to current earthquake information. It includes a link to viewing seismographs from the latest worldwide or local earthquakes, up to date information on recent earthquake activity and maps from the National Earthquake Center. There is also a link to Resources for K–12 teachers. However, when I clicked on that link the Web site was unavailable. The Seminars page lists seminars available at Cal Tech. Seismolife links you to a page that discusses life at the Seismological Lab. There are links to places to visit in the surrounding areas. Again this is useful information if you plan to go there. The People page is a list of faculty, staff, and students. The Research page lists the areas of research carried out at the Seismological Lab. Clicking on these areas takes you to the research. There is a link at the bottom to Cal Tech's preprints collection, which is an "online database of aca-

demic preprints, papers submitted for publication but not yet published." The entire archive can be browsed or specific searches can be made. Preprints can be downloaded in Adobe PDF. The Late Breaking link contains current press releases. This Web site contains some useful information that might be useful to teachers or researchers. If you bypass the topics specifically related to Cal Tech, the site warrants a quick browse.

Soil Science

Title: Soil Science
URL: http://www.rdg.ac.uk/acadepts/as/ugrad.html
Grade Level: High school
Best Search Engine: www.google.com
Key Search Words: soil sciences
Review: This site is the Department of Soil Sciences at the University of
    Reading in the United Kingdom. The home page has links for re-
    search, courses, contacts, facilities, events, news, and other links. The
    research area has a list of Ph.D. topics. The four research teams are
    Soil Vegetation Atmospheres Transfers, Sampling and Spatial Analy-
    sis Research Team, Soil and Water Quality Team, and Soil Plant Mi-
    crobe Interaction Team. Each team page explains the purpose of each
    team and its research. There are not many graphics or much interac-
    tive work, but the studies seem to be interesting. I have a cousin who
    is going to college to study turf management, and I am sure he would
    be able to use a site like this.

Title: The Soil Makers
URL: http://www.earthlife.net/insects/soileco.html
Grade Level: Middle school through high school
Best Search Engine: www.yahoo.com
Key Search Words: soil & science
Review: This is a very interesting site that provides information about the
    creation of soil. This is basically a text site that discusses insects and
    micro-organisms that produce soil. The site begins with an introduc-
    tion called "Soil Zoology," which includes information about decom-
    posers, predators, and secondary decomposers. A section about slurry
    includes information about soil organisms and organic manures. De-
    composition and the importance of arthropods in soil creation is also
    discussed. This site lacks the bell and whistles of other Web sites, but

it is one of the few sites that provides useful information about the science of soil. References are provided to enable the user to locate additional information on a related topic.

Title: Natural Resources Conservation Service (NRCS)
URL: http://www.nrcs.usda.gov/feature/education
Grade Level: Grades K–8
Best Search Engine: www.mamma.com
Key Search Words: soil science
Review: This Web site is an outstanding resource for teachers and parents. The Web site was developed by the U.S. Department of Agriculture. The site offers a variety of resources for lesson plans and projects. You can look up articles on soil and soil science. Teachers can also access lesson plans on many topics regarding soil. The site offers a search area where you can search for an enormous amount of information. In addition the site includes information on types of soil, maps, pictures, and links to resources. NRCS has even compiled a list of books that are appropriate for a variety of students. Soil is just one small area of the Web site. It is devoted to information on agriculture. Students can research a number of projects. The site is easy to navigate and well maintained. I highly recommend this site.

Title: Soil Science Education
URL: http://ltpwww.gsfc.nasa.gov/globe
Grade Level: Grades 3–5
Best Search Engine: www.google.com
Key Search Words: soil science
Review: This Web site was one of my favorites! The home page alone is very colorful and engaging. There were numerous photos, graphs, and charts. The layout was user friendly. There were also many categories of information to choose from. In the What's new category there was a soil gallery with colored photos of many types of soil. The Features section had a great deal of information. There were also links, resources, soil science basics, soil and society, soil and the environment, soil and students, soil and agriculture, and a place to search for information that wasn't there. Whether you are teaching or learning about soil, I think looking at this site would be an excellent place to begin.

Title: Applied Soil Science
URL: http://www.members.aon.at/soil
Grade Level: High school

Best Search Engine: www.dogpile.com
Key Search Words: soil science
Review: Wow, something I didn't know, topsoil is one of the most complex mediums on the planet. After much searching, I found this site, which has more than a dozen sections. They include: Web site changes, Who thinks about, Offensive smell reduction, Calibration Free Analysis, and Horticulture. The site then offers services such as database, sampling, composting, quality assurance & control, soil assessment, and environmental compatibility study. The site also has links to curriculum, references, and products. I didn't think this was the best site but most of the others were membership driven and newspaper articles about soil.

Title: Soil Science Society
URL: http://www.soils.org/sssagloss
Grade Level: Middle school and above
Best Search Engine: www.google.com
Key Search Words: soil science
Review: This Web site is offered by the Soil Science Society of America. It is an Internet glossary of Soil Science terms. It is an excellent site to access for the proper definitions of terminology in this area of science. One can access this information in a number of ways. "Search" is used to type in a term for its definition. Using "Browse" you can click on a letter of the alphabet and then click on the term you need when you identify it. Under the letter "C" there are 183 terms listed; under the letter "S" there are 312. As you can see, this glossary is quite extensive. Most useful to the advanced high school student are the tables, figures, and appendices offering SI and non-SI unit conversion factors. All of these tables, figures, and appendices are printable for future reference.

Tornadoes

Title: Tornadoes
URL: http://www.usatoday.com/weather/tg/wtornwhat.html
Grade Level: Grades 6–8
Best Search Engine: www.dogpile.com
Key Search Word: tornadoes
Review: This is a multipage Web site with many links to other areas and terminology as it relates to tornadoes. I do not recommend this site as a lesson; however, it is an excellent supplement to a lesson on torna-

does. It covers several areas of importance with real-time links for video. Some of the areas covered include microbursts, large hail, upper level disturbance, wind shear, thunderstorms, and supercells. Each term has a link for further research to more clearly define a tornado and the differences between terms and common confusion between each. One of the unique things about this Web site is that there is a feedback link to evaluate the article and information.

Title: How Tornadoes Work
URL: http://www.howstuffworks.com/tornado1.htm
Grade Level: Grades 5–8
Best Search Engine: www.yahoo.com
Key Search Word: tornadoes
Review: This page, part of the HowStuffWorks Web site, has some great illustrations and explanations of tornadoes. One page describes how a vortex forms on a molecular level. Another page provides a chart of the Fujita Scale with a link to damage descriptions for each level tornado. This site strikes me as a great resource for teachers or parents who want to answer the kinds of questions often posed by youngsters.

Title: The Tornado Project Online
URL: http://www.tornadoproject.com
Grade Level: Grades 6 and up
Best Search Engine: www.metacrawler.com
Key Search Words: tornado education
Review: This site is mainly an archive of every notable tornado that has ever occurred. It's collection of tornadoes dates back as far as the 1700s and is sorted by year, state, country, or continent. It touches upon tornado safety, tornado stories, and the Fujita Scale (a tool for measuring magnitude of tornadoes). If you are researching past tornadoes, this is the site, but if you are trying to teach the science behind tornado development, this may not be the site (or maybe it is, I didn't have time to navigate the entire site). There are also links to other good sites that the creator of this site found to be related and interesting. I checked a few out myself and loved http://www.jimreedphoto.com, which contained dozens of excellent severe weather photographs.

Title: Tornadoes
URL: www.chaseday.com/tornadoes5.htm
Grade Level: High school

Best Search Engine: www.google.com

Search Word: tornado

Review: As you navigate into the site you come across color pictures of tornadoes. There are six parts to this site. With each of the pictures is a detailed description of what it looks like and what happens during these types of tornadoes. Each picture is a real tornado that has occurred in the Midwest. The site talks about how the funnel does damage and when the condensation funnel is not on the ground. It gives information on half funnels and classic tornadoes. This link allows you to see lightning conditions and dissipating tornadoes. The rope out and the thinnest rope talk about how a tornado ends. The suction spot and parasite vortices talks about two funnels (suction spots) that are rotating around a larger tornado. Part 4 talks about two tornadoes at once and the HP supercell dilemma. Part 5 discusses how meteorologists need to see the tornado and the clouds. It talks about the categories and the shearing jet stream winds. It gives information on safety and tornado chasers. The last part discusses large cone, long snake and straight tube tornadoes and the damage they can cause. The site allows you to click on other links to give you more information. I found this site interesting because of the pictures.

Title: Tornadoes

URL: http://www.fema.gov/kids/tornado.htm

Grade Level: Grades 2–4

Best Search Engine: www.msn.com

Key Search Words: tornadoes and kids

Description: This site was created by the Federal Emergency Management Agency (FEMA) for kids. The first page of the site briefly describes tornadoes and explains the difference between a tornado watch and a tornado warning. The site allows students to click on links that provide information on tornado safe rooms, the Fujita Tornado Scale, disaster intensity scales, and how to keep yourself safe when a tornado strikes. The site also provides tornado videos and interactive games for the students. A neat part of this site is that it displays student projects that deal with tornadoes. The site is colorful and interesting. There are great graphics and photos that help students to understand information about tornadoes. The information is clear and easy for lower elementary students to understand. Students can e-mail FEMA with questions and comments too. I would recommend this site to anyone teaching a unit on tornadoes.

Title: USA Today.com—Resources: Understanding Tornadoes
URL: http://www.usatoay.com/weather/tornado/wtwistO.htm
Grade Level: Upper elementary and above
Best Search Engine: www.metacrawler.com
Key Search Word: tornadoes
Review: This Web site was put together by *USA Today*. The home page
provides links to detailed studies of the tornado. It is divided into 10
main topics: tornado basics; U.S. tornado climatology; tornado news,
history, and information; smaller vortices, including waterspouts; tornado photos, movies, and videos; stalking twisters—tornado chasing;
tornado research; questions and answers about tornadoes; Skywarn,
the nation's volunteer storm spotter network; and On the Web. Each
of these topics is further divided into subtopics that link you to more
detailed information about the topic. Clicking on the link to the
Web gives additional sites to go to for tornado information. In addition to information about tornadoes there are links to information on
other types of severe weather (lightning, thunderstorms, floods and
droughts, hurricanes, and winter weather). This is a site with a wealth
of information in an easy-to-use, easy-to-read format. If you can stand
the pop-up ads, it's a great site to explore.

Title: Tornadoes
URL: http://www.uscase.edu
Grade Level: Grades 4 and up
Best Search Engine: www.dogpile.com
Key Search Words: tornadoes
Review: This site I liked because it is geared toward children. It starts out
talking about the misconception that *The Wizard of Oz* movie had
left in the mind of children—that tornadoes lift the house and land it
in a fairy-tale place. This site goes on to explain exactly what a tornado is and why it is so destructive. The site also has links that allow
the teacher to research additional information for classroom use.

Title: Tornadoes, Going Around in Circles
URL: http://whyfiles.org/013tornado/index.html
Grade Level: Elementary school
Best Search Engine: www.yahooligans.com
Key Search Word: tornadoes
Review: The main page opens with an account of what happened during
a tornado in May 1999 in Kansas. After the description, there are
seven questions about tornadoes. When you click on the question,

you will be taken to another page that provides the answers. When you click on What are tornadoes, you are taken to a page that features a colorful map showing Tornado Alley and the causes of tornadoes. There is also a detailed description of what causes tornadoes. Other links provide the answers to Where does a tornado get its energy from?, What's the latest word on tornado prediction?, What do you do during a tornado warning?, and How does a tornado affect the land? There is also an interactive quiz to test your tornado knowledge. At the end, there are links provided to the creators of the Web site. This is a great site for elementary students. The graphics and information are easy to understand.

Tsunamis

Title: Wind
URL: http://www.pmel.noaa.gov/tsunami-hazard/kids.html
Grade Level: Pre–K and up
Best Search Engine: www.ixquick.com
Key Search Word: tsunami
Review: The main Web site features the headlines: The National Tsunami Hazard Mitigation Program: Links for kids. There are approximately nine links for kids: Print out and enjoy the Tsunami Trivia game; NGDC Kids hazard page; Tommy Tsunami Coloring book; FEMA's Tsunami page; Tsunami: Great waves; PBS Savage Earth; Tsunami word search game; Tsunami text books for preschool through high school; and the Washington Emergency Management Division kids page. I found all the subsites helpful and interactive. I also discovered the coloring book is fun to color and seemingly accurate. The authors of this educational Web site are from the United States Department of Commerce. All the links for kids are age appropriate for children in preschool and I found the main Web site helpful and interactive for children 3 years old and up. This is also a beneficial Web site for teachers. I also found that the entire Web site's information can be printed because the author provides permission.

Title: Tsunamis
URL: www.geopys.washington.edu/tsunamis/welcome.html
Grade Level: Grades 6–12
Best Search Engine: www.ask.com
Key Search Word: tsunami

Review: This is a World Wide Web site that has been developed to provide general information about tsunamis. This site was developed with a broad audience in mind. It contains extensive background information that is intended primarily for the general public, including information about the mechanisms of tsunami generation and propagation, the impact of tsunamis on mankind, and the tsunami warning system. The site also contains more detailed information about recent tsunami events that will be of interest to students doing tsunami research. This site is efficient for the research needed in any high school science class. The site itself is not that dull, but it is very scientific in its content and could become boring to the average student.

Title: Hydrology Investigation
URL: http://www.pbs.org/wnet/savageearth/tsunami/index.html
Grade Level: Grades 4–6 (animation for lower grades)
Best Search Engine: www.google.com
Key Search Word: tsunami
Review: This subsite is part of the PBS Savage Earth site, which provides much information about Earth Science. The introductory page on tsunamis starts with a general introduction by way of an article called "Surf's Up." There is a video link on this page entitled "Tsunami Survivor," which you can view with QuickTime. (If you do not have QuickTime on your computer, you can download the free program though a link on the site.) The link to sidebar 1 at the top of the page discusses early warning systems, a computer model of a tsunami, and photos of the aftermath of a tsunami. There is also a Flash animation of a tsunami spread. The link to sidebar 2, entitled "Remembrances of Waves Past," discusses how scientists can learn about tsunamis that occurred prior to the modern scientific techniques through studying the soil left behind. The animation link takes you through a series of animations about the cause and effects of a tsunami. This is a very graphic way for students to understand the cause and effects of a tsunami.

Title: Welcome to Tsunami!
URL: http://www.geophysics.washington.edu/tsunami/welcome.html
Grade Level: High school and teachers
Best Search Engine: www.google.com
Key Search Word: tsunami
Review: Welcome to Tsunami! is an interactive information resource about tsunamis. The Web site was developed to provide general in-

formation about tsunamis. The site reaches out to a broad audience and is well presented and organized. It contains extensive background information about the mechanisms of tsunami generation and propagation, the impact of tsunamis on humankind and the Tsunami Warning System. It also contains information about recent tsunami events that tsunami researchers might find of interest. The table of contents allows one to access general information related to tsunamis, survey and research information, miscellaneous information such as related sites, and background information on the development of this site. The site is user friendly and certainly of value to anyone who would like to expand their knowledge related to tsunamis.

Title: Tsunami—The Big Waves
URL: http://observe.arc.nasa.gov/nasa/exhibits/tsunami/tsun_physics.
    html
Grade Level: Middle and high school
Best Search Engine: www.yahoo.com
Key Search Words: tsunami & earth science
Review: This was a great site for students and teachers. This site provided
    a description of what tsunamis are, pictures of the damage left behind
    after a tsunami, and discussed the causes of tsunamis. Diagrams
    showed a comparison of wind-generated waves and tsunami waves
    created deep under water. It provides information about wave speeds
    and what happens as the wave approaches the shore. Detection sys-
    tems are discussed, and tips are provided as to what actions to take
    during a tsunami warning. Toward the end of the site there is a map of
    the areas of the world effected by tsunamis. This site also included a
    tsunami quiz and word search. This site was easy to navigate and pro-
    vided good information for students.

Title: Welcome to Tsunamis
URL: http://Search.yahooligans.com/search/ligans?p=tsunamis
Grade Level: Grades 6–8
Best Search Engine: www.yahoo.com
Key Search Word: tsunamis
Review: This is an interactive tsunamis resource site. The site is hosted by
    the Geophysics Department at the University of Washington. It is
    broken down under the table of contents: General tsunami informa-
    tion, Alaska Tsunami Warning Center, the physics of tsunamis, A sur-
    vey of great tsunamis, the tsunami warning system, tsunami hazard

mitigation, Tsunami survey and research information, Recent tsunami events, and Tsunami research, and Miscellaneous information. Throughout the site it talks about famous tsunamis in history.

Title: Tsunamis & Earthquake
URL: http://walrus.wr.usgs.gov/tsunami/
Grade Level: High school
Best Search Engine: www.mamma.com
Key Search Word: tsunamis
Review: This Web site was developed by the U.S. Geological Survey, which has the motto, Science for a Changing World. The Web site offers information on numerous topics. I concentrated mainly on earthquakes and tsunamis. The Web site is best suited for older students and high school aged students. The site offers links to many other Web sites that provide information on tsunamis. The page begins by defining tsunamis and explaining how they are created. There are examples that include pictures and explanations. In addition students can access the site to see virtual realty tsunamis. The animations are great. This site is a great resource for teachers and students. The information available is a little technical. I recommend this site for researching projects and developing lesson plans.

Volcanoes

Title: Volcanoes
URL: http://www.geol.ucsb.edu/~fisher
Grade Level: Middle and high school
Best Search Engine: www.metacrawler.com
Key Search Word: volcanoes
Review: This site was informational and was broken down into categories. Some of the categories were Gasses, Eruptions, Forms, Flows, and Fallout. Also, there were lists of associations, references, organizations, and institutions for study. This site was updated by the University of California at Santa Barbra by Professor Richard Fisher.

Title: Volcanoes
URL: http://www.geology.sdsu.edu/how_volcanoes_work/
Grade Level: Grades 1–12
Best Search Engine: www.37.com
Key Search Word: volcanoes

Review: An excellent Web site with QuickTime animations. The site is broken down into several categories: Eruption dynamics, Volcano landforms, Eruption products, Historical eruptions, and Planetary volcanoes. There is also a crossword puzzle, plus related links. This is a very extensive Web site good for high school students. It allows for study in many areas of volcanic activity and can be implemented in many ways to keep a class active. It has an evaluation page for rating the article and providing feedback. It also provides a link to the department of geological sciences, the site provider.

Title: Volcano Lovers
URL: http://whyfiles.org/031volcano
Grade Level: Grades 4–8
Best Search Engine: www.metacrawler.com
Key Search Words: volcanoes education
Review: Very good site to be used as an introduction to a unit on volcanoes. A couple of days could be spent on this site, or quickly in one day. This site would get students excited about volcanoes. It's concise and to the point. It dedicates one page each to famous volcanoes, the basic science behind their formation and creation, forecasting, earth-shaping effects and physical phenomena, and biological impact. It also contains some great historical events like the loudest volcanic explosion ever, or the first people to witness the birth of a volcano. Overall great site to get students excited about learning about volcanoes.

Title: Volcanoworld Alternate
URL: http://www.volcanoworld.org
Grade Level: Elementary school
Best Search Engine: www.webcrawler.com
Key Search Word: volcano
Review: I wish I had known of this Web site last year when I taught landforms to my third-grade ESL students! This site contains everything an elementary teacher could want in a science Web site. It has legends, excellent pictures, lessons, maps, definitions, a store, and probably most importantly, an index. There's a volcano calendar, which allows you to see if there were eruptions on any given day. You can search for volcanic activity/information by region, country, name, or descriptions. There is a page that lists volcanic activity around the world in chronological order, from most recent on back. It is an all

around useful site and very user friendly. It is maintained by the University of North Dakota and has links to the U.S. Geological Survey site on volcanic activity. This is a good one!

Title: Volcanoes Online
URL: http://library.thinkquest.org/17457/english.html
Grade Level: High school and above
Best Search Engine: www.google.com
Key Search Word: volcanoes
Review: As you enter the site you get a black screen with a big volcano picture. On this page you have a main menu that includes Plate tectonics, Volcanoes, Volcanic database, Games, Comics, Teach, Top sites, and About us. On this page you also see a cartoon type world and a caption. As you navigate into each area you get a table of contents. Under Plate tectonics you will see pictures and detailed descriptions on the continental drift, sea floor, spreading, subduction and other information. The Volcano link discusses hot spots, types of eruptions, how it erupts, hazards, power of volcanoes, and features. Within each link you will see black and white photos with detailed information along with comics. The database has information from around the world on volcanoes. You can search by continents and are able to view 38 volcanoes and read about 68 other volcanoes. The game section allows you to play volcano games while testing your knowledge of volcanoes. The comic link has Galvin in a classroom discussing volcanoes and captions. The teach link allows you to use the Web site in the classroom. You are able to download lesson plans and share lessons with the site. The about us discusses who designed the site. Three teenagers from three countries compile information, graphics, and research to develop the Web site. The site is made for high school and above. The site gives a lot of information and detailed description of volcanoes.

Title: How Volcanoes Work
URL: http://www.howstuffworks.com/volcano.htm
Grade Level: Grades 5–8
Best Search Engine: www.yahoo.com
Key Search Word: volcanoes
Review: This is another HowStuffWorks Web site. The site includes a nice explanation of the process by which volcanoes are created. There are illustrations, photographs and, my favorite, animated illustrations that go with the text explanations. There are plenty of links

to other related pages within the site. This is a site that would be very useful when introducing middle school students to the study of volcanoes.

Title: How Volcanoes Work
URL: http://www.geology.sdsu.edu/how_volcanoes_work/
Grade Level: College and teachers
Best Search Engine: www.metacrawler.com
Key Search Word: volcanoes
Review: This Web site was constructed by Dr. Vic Camp of the Department of Geological Sciences at San Diego State University. It is sponsored by NASA under the auspices of Project Alert (Augmented Learning Environment and Renewable Teaching). The home page explains that the information in each section builds on previous sections so that those less knowledgeable in the subject matter should go through it progressively whereas those with greater knowledge can navigate it as their interests dictate. The home page contains nine main headings: Eruption Dynamics, Volcano Landforms, Eruption Products, Eruption Types, Historical Eruptions, Planetary Volcanism, Volcano Crossword, Volcano Links, and Awards. Each of the main headings is divided into subtopics with additional information. At the end of most of the sections there is a test so you can see what you have learned. There is only one volcano crossword. The Volcano Links takes you to a site that has a wealth of information. There are numerous educational links. There are links to volcanoes of the world (divided by region) and there are virtual field trips. If you need to find a volcanologist you can find a link to that too. The Web site contains over 250 images with explanations about what you are viewing. This site is definitely worth exploring.

Title: Volcanoes
URL: http://www.fema.gov/kids/volcano.htm
Grade Level: Grades 2–4
Best Search Engine: www.msn.com
Key Search Words: volcanoes and kids
Description: This site was developed by the Federal Emergency Management Agency (FEMA) for kids. The site is a great resource for elementary students. The site clearly describes a volcano and gives many facts about volcanoes. The site gives students opportunities to click on links to find out about mapping new lava and to learn about disaster intensity scales. The site also has a link so the students can learn

about Mount St. Helens. Students are able to find out how to stay safe from volcanoes and they can view great pictures of erupting volcanoes. The students will love this bright and interesting site. It is perfect for children being introduced to volcanoes!

Title: Volcano Expedition from the Field in Costa Rica
URL: http://www.sio.ucsd.edu/volcano
Grade Level: Elementary and middle school
Best Search Engine: www.dogpile.com
Key Search Word: volcanoes
Review: There are five main links on this site: The Expedition, Volatiles and Volcanoes, the People, In the Lab, and Volcano Q & A. Visitors to the site will read about the research findings, watch videos of the scientists and view pictures of the volcano's surroundings. You can also find out about the latest volcano activity by clicking on the large picture of the volcano. You will then be taken to another page that gives a picture and information about the volcano and its latest eruptions. Difficult words are used throughout the selection and/or pictures. Volatiles and Volcanoes offers a detailed introduction about volcanoes, and continues on to discuss volatiles, isotopes and geochemical tracers. By clicking on The People, you are able to contact the science team and exploration team. Information about their qualifications and experiences are included in a detailed summary done by the actual person. In the Lab details experimental findings of several of the scientists. Topics include geothermal fluids, lavas, and mass spectrometry. Volcano Q&A provides easy to understand answers to several questions related to volcanoes, such as What is lava? How is a volcano made?, and How long do volcanoes last?. The name and grade level of the school are provided. The answers are brief and to the point, and on the grade level of the person who asked the question. There are also Q&As related to the Expeditions, Safety/Danger, Advanced topics, and Technical/Web problems. Back on the main page, links to the volcanoes are featured. Each link provides a vivid picture of the volcano, as well as a brief description about it and any recent activity. There is also a link to a map of Costa Rica. Web site visitors can click on the map and see an enlarged version of the map which includes the locations visited during the expedition. This is a good site. It is not too scientific. Teachers and students will enjoy the great images of volcanoes. The Q&A section is very informative, as it includes "real" questions that kids actually ask.

Title: How Volcanoes Work
URL: www.geology.edsu.edu
Grade Level: College
Best Search Engine: www.googple.com
Key Search Word: volcanoes
Review: This site is user friendly in that it is easy to navigate. The site begins dealing with volcanoes and how they are formed. On the left side
of the screen are several interesting links. Each link has the main
heading with subheading below. The main topics are Eruption dynamics, Volcanic landforms, Eruption products (for example, lava
flow types), Eruption types, Historical eruptions, Volcanoes in other
worlds, Volcano crossword, Volcano links, and Awards. The site also
contains images and animations, which are really neat to look at as
they show exactly how eruption occur. I think you will enjoy this site.

Title: Global Volcanism Program
URL: http://rathbun.si.edu/gvp
Grade Level: Grades 7–12, college
Best Search Engine: www.yahoo.com
Key Search Word: volcanoes
Review: This site gives a lot of information about specific volcanoes.
There is only a brief section answering commonly asked questions,
such as What is a volcano?, What is an eruption?, Where are the active volcanoes?, and What is the highest volcano?. The other sections for this site focus on specific volcanoes. You can do a volcano
search based on location or name. You can choose to get a one-line
summary or an in-depth review of that particular volcano. There is
also an entire section focused on recent volcanic activity. They have
weekly reports on volcanic activity and monthly bulletins so you can
stay up to date. This site is not the best site, but there is a lot of information for those studying specific volcanoes and those wanting to
know when there has been any activity.

Title: Plate Tectonics
URL: www.volcano.und.edu
Grade Level: Educators
Best Search Engine: www.dogpile.com
Key Search Word: volcanoes
Review: I was drawn to this site because it specifically mentions how it is
teacher friendly. I found the information quite informative. The site

begins with an introduction to its site. The opening page is only the introduction. As you click on one of the three links at the bottom of the page—Teacher's guide, Volcano world, or Next—you will find new and exciting information. The teacher guide has photos and additional links for further information. The volcano world link takes you to a page hosted by the University of North Dakota. This link features many important news fact about volcanoes. Finally the next bottom takes the viewer to additional pages that show the earth below with readable information explaining the picture being viewed. The continuous pictures describe continental drifts, continental drift fossils, continental drift glaciation, location of continental drifts, and types of continental drifts. You will really enjoy viewing this site and learning from it.

Water

Title: Water Environment Federation
URL: http://www.wef.org
Grade Level: Grades 6–12
Best Search Engine: www.ask.com
Key Search Word: water
Review: The WEF is the Water Environment Federation. The WEF is a not-for-profit technical and educational organization. Its vision is to preserve and enhance the global water environment. The federation's focus is not only point sources of pollution (pollution caused by direct sewage discharge into public waters), but nonpoint sources of pollution (pollution caused by indirect contamination of water, for instance, runoff of fertilizers into rivers after storms). This site provides a range of materials describing today's water quality issues, including household hazardous waste, biosolids recycling, and watershed management. WEF also works to inform public officials and the media about water quality through the Web site. This site is not only one for informational purposes due to researching but also it is something that more and more people should be aware of. The WEF not only informs the public of water pollution, but also is trying to do something about it through the federation's membership. This is a good site for high school students to become involved in.

Title: Water Science for Schools: All About Water
URL: http//ga.water.usgs.gov/edu.html
Grade Level: Elementary and middle school

Best Search Engine: www.metacrawler.com

Key Search Words: water, water science

Review: This Web site was developed by the highly credible U.S. Geological Survey, part of the U.S. Department of the Interior. Addresses and contact information are listed right on the home page of the site. Pictures, maps, and interactives are used to disseminate information, take comments and opinions, and test student's water knowledge. Each main water topic, such as Earth's Water, Water Basics, Water-Use Information, Special Water Topics (acid rain, saline water, water quality), has its own rubric of study geared to classroom use. The topic areas use cute graphics, which could possibly pique children's interest. Additional USGS information is available through a link. There are also features of help, search, glossary of terms, Q&A, picture gallery, certificate of completion, and fascinating links to nationwide schools and groups who are also involved with water science projects. This site could be a refreshing addition to any school science class.

Title: Water

URL: http://ncsu.edu/sciencejunction/depot/experiments/water/kids.html

Grade Level: Grade 3 and up

Best Search Engine: www.ixquick.com

Key Search Word: water

Review: The main Web site features the headline Water-What ifs for kids. This main page features 11 sections to explore. The author of this main Web site is April J. Cleveland for Science, located at North Carolina State University. I explored two subsites called What is water and What's wrong with this picture? What is water features three headlines that are user friendly. The three headlines are Hydrogen and oxygen creates water, Three phases to a quick change, and Forming the link to life. This subsite also has a diagram of two hydrogen and one oxygen to describe a water molecule. What's wrong with this picture seems like a fun and interactive way to find things that are wrong with the environment. The caption emphasizes that people are taking care of their home and car. A student can click on the various parts of the picture and get a surprise. The surprise is an explanation of why the environment should be protected and not damaged. I found the main Web site helpful and interactive for children 8 years old and up. This is also a beneficial site for teachers. The What's wrong with this picture can be displayed on a overhead projector and whole-class discussion can be formed.

Title: Drinking Water for Kids
URL: http://www.epa.gov/OGWDW/Kids/cyde.html
Grade Level: K–12
Best Search Engine: www.yahoo.com
Key Search Words: water & earth science
Review: This site is part of the Environmental Protection Agency Web
    site and developed for the purpose of educating children about
    ground water. This site provides games and online activities for the
    kindergarten through sixth grade and for seventh through 12th grade.
    There are also classroom activities and experiments for K–6 and 7–12
    grade levels. A section about Kid's Health discusses the important
    role water plays in keeping children healthy. There are teacher activ-
    ities and student activities. In addition they provide a page of links
    for older students to find additional information they may be inter-
    ested in or need for a report. Some of the activities for the kids to try
    include a diagram of a water treatment facility with a word scramble.
    There are also drinking water bloopers, a word scramble with a chart
    of how much water we use during everyday activities, like taking a
    shower. After reviewing this site it was obvious that there were two
    goals. One was to raise awareness in children about water pollution.
    The other was to increase awareness about the amount of water we
    use and waste each day. This is a good educational site for a wide va-
    riety of age levels.

Title: Water in the City
URL: http://www.fi.edu/city/water
Grade Level: Middle school
Best Search Engine: www.yahoolligans.com
Key Search Word: water
Review: The first thing one notices about this site are the beautiful graph-
    ics. The site begins by asking a user to brainstorm about water. Great
    idea. From there it goes into the usual items one would normally find
    in this type of site. Water basics, Water science, Philadelphia water-
    ways (the site is hosted by the Franklin Institute), Worldwide water-
    ways, and Reference and activities. I am always interested in the
    interactive pages. This site doesn't disappoint. It starts with your
    basic word search, EPA Office of Water kids stuff, Water contamina-
    tion experiment, Build your own water cycle, Build your own aquifer,
    Water filtration, and Groundwater guardian. I like the fact that kids
    can build their own water experiments.

Title: Water Education and Kids Corner
URL: http://www.mojavewater.org/Mwa800.htm
Grade Level: Grades 2–4
Best Search Engine: www.google.com
Key Search Words: water education
Review: This Web site was developed for a water education project. Much of the Web site was devoted to the program. There was a Mojave Water Association calendar listing all upcoming meetings. There were also MWA projects described. The part of the Web site that I found interesting was the Kids Corner. In the Kids Corner, you could find water conservation fact and tips, what groundwater is, and the water cycle. There was also a section of games and activities. However, this part was still under construction. There was only a word search. I thought this site might be a good starting point in a search for information about water.

Title: U.S. Geological Survey (USGS) Water Resources of the United States
URL: http://water.usgs.gov
Grade Level: Elementary through college
Best Search Engine: www.metacrawler.com
Key Search Word: water
Review: This extremely comprehensive Web site offers a plethora of information. It is a government site maintained by the Office of Surface Water. Overall, it is a rather advanced, sophisticated site but it can be used by someone as young as the elementary level if one accesses the area on Educational Resources. Here you will find an area called "Frogweb"; an area for educational posters; an area called "Water Science for Schools"; and the USGS Learning Web. There is information on government projects and programs, fact sheets (arranged according to themes), abstracts for research, news releases covering the latest USGS news, and listservs. There is a section dedicated strictly to topics such as Acid Rain, Waterwatch, Water Use, and Water Quality. I found the section showing the latest information on the Drought Watch to be interesting. This area offers maps indicating areas of the country affected by the drought and can be viewed from a daily or monthly perspective across the United States. There are maps and graphs of water conditions that can be viewed state by state. As I stated in the beginning of this review, the information is endless and it covers an incredible variety of topics for a multitude of

levels. I highly recommend this site as a great source of information for many levels of learners.

Waves

Title: Waves
URL: http://www.seafriends.org.nz/oceano/waves.htm
Grade Level: High school
Best Search Engine: www.webcrawler.com
Key Search Words: water and waves
Summary: This site is maintained by Dr. Floor Anthony and the Sea Friends' Marine Conservation & Education Center in Leigh, New Zealand. It is comprehensive in its approach to the subject matter. It has every definition you could possibly need when referring to ocean waves. It is chock full of graphs, photos, maps, and information relating to waves. There are pages about how earthquakes have created giant waves, or tsunamis. There are pages on tides and currents. The good doctor even has a Q&A page with a little picture of himself smiling at the top. There are not many links, but as I said, the site is comprehensive in itself.

Title: Kettering University
URL: http://www.gmi.edu/~drussell/Demos.html
Grade Level: All grade levels
Best Search Engine: www.metacrawler.com
Key Search Word: waves
Review: This site also gave me some ideas for my science fair at school. It is distributed by Kettering University and is updated frequently by professors at the university. There is also a link to the professor's home page as well as private e-mail. There is a link to a School Zone education series that houses information on many science topics as well as other elementary areas of learning.

Title: Waves Science/Physics
URL: http://www.askeric.org
Grade Level: Grades 3–5
Best Search Engine: www.metacrawler.com
Key Search Word: waves
Review: This is a good Web site for educators. It is presented in lesson-plan format with three options of length depending on class time. Two major areas are covered as they relate to waves, crest, and

trough. Hands-on lab activities are also given with resources for materials and procedures. Book links are offered at the end, something I have not seen in previous site reviews. The book links seen very well researched. The article was posted in May 1994.

Title: All about Oceans and Seas
URL: http://www.enchantedlearning.com/subjects/ocean/waves.shtml
Grade Level: Elementary, teachers of elementary science
Best Search Engine: www.metacrawler.com
Key Search Words: ocean waves
Review: This Web site was created by Enchanted Learning. At the top of the home page there are links to 11 topics: Introduction, What Causes Waves?, Why Is the Ocean salty?, The Water Cycle, Ocean Animal Printouts, Ocean Crafts, What Causes Tides, Why Is the Ocean Blue?, Undersea Explorers, Coral Reefs, and Intertidal Zones. The introduction provides an outline of the Earth's oceans and is easily understood. There is a chart that lists the area, average depth, and deepest depth of each ocean. Clicking on the links to the topics brings you to information related to that topic and additional Web links. One of my favorites was Ocean Animal Printouts, which contains lists of ocean animals with a picture. When you click on the picture you come to a page with an article and picture that can be printed out. I also liked Ocean Crafts because it contains interesting projects to do with children. This site is easy to navigate with great information and ideas to use with kids. It is definitely worth your time, especially if you teach marine science.

Title: The Birth of a Wave
URL: http://geography.about.com/gi/dynamic/offsite.htm?site=
    http%3A%2F%2Flibrary.thinkquest.org%2F2804%2Fwave.html
Grade Level: AP high school and college
Best Search Engine: www.metacrawler.com
Key Search Words: ocean waves
Review: This site gives a great description of ocean waves. At the search window, type in "ocean waves." It first gives the definition of a wave and then it describes the three types of waves: longitudinal wave, transverse wave, and orbital wave. The site also gives clear information on all the parts of a wave (trough, crest, wavelength, height, period, frequency, and amplitude). The students can then read about the difference between shallow water waves, deep water waves, and wind-generated waves. The site gives some graphics to help students

understand the information about the movement and parts of ocean waves. This site is difficult to understand and is recommended for AP high school and/or college students.

Title: Parts of a Wave
URL: http://id.mind.net/~zona/mstm/physics/waves/partsOfAWave/
    waveParts.htm
Grade Level: High school and college
Best Search Engine: www.yahooligans.com
Key Search Words: ocean waves
Review: This Web page is part of the Zona Land Web site. There are nine links on the main page: Introduction, Parts of a Wave, Transverse and Longitudinal Waves, Wave Adder, Wave Propagation and Huygen's Principle, Interference, Standing Waves, 3-D Surface Waves, and General Problems. The Introduction provides a detailed overview of what a mechanical wave is. Mechanical waves are described as water and sound waves. There is a diagram showing the different parts of waves and descriptions of each. There is an interactive picture that illustrates amplitude, wavelength, and phase shift. Parts of a Wave includes a link to each of the different parts of a wave, crest, trough, amplitude and wavelength. The definitions and pictures are clear and easy to understand. There is also an animation that illustrates the concept of wave frequency. Transverse and Longitudinal waves feature animated pages that help to illustrate the written definition of each type of wave. The link to Wave Adder includes another interactive diagram which shows what happens when two waves interfere. The instructions on how to manipulate the animation are clearly stated and very easy to follow. Wave Propagation and Huygens' Principle offers three links, Single Source Patterns, Several Source Patterns, and Almost Continuous source patterns. Each link leads the Web site visitor to another page that features an animation that describes each topic. The Interference Link features what I thought was a very confusing picture of a wave interference pattern. There are links to other pages, such as constructive and destructive interference and different types of wave interference. The pages also feature an interactive diagram which shows such interference. Again, the instructions for the animation are easy to understand. Wave Reflection gives a brief definition of the term and includes two links to other sites about fixed-end and open-end wave reflection. A slide show helps the Web site visitor learn about fixed- and open-end wave reflection. The Standing Waves page gives detailed information and provides several links to related sites, most of which feature inter-

active diagrams. Formulas for calculating the velocity of a wave are featured on the Three Dimensional Wave page. Diagrams and interactive animations are also featured. General Problems features a quiz of sorts with drop-down answers to FAQs about waves. This is a great site for science-minded people. I was able to comprehend only the very basics, such as parts of a wave. It would be a good site for high school and college students who are learning about waves. The interactive animations offer a great hands-on experience.

Title: Ocean Waves
URL: http://www.geography4kids.com
Grade Level: Grade 5 and up
Best Search Engine: www.google.com
Key Search Words: ocean waves
Review: This Web site is designed to inform the reader about different sciences. The site has links that are colorful and fun to look at. The links include Biosphere, Beach Drainage, Dune Formation, Growth Phases, Plant Distribution, Steam Mouth, and Tidal Flat. The page opens with a Natural Occurrence section, Information, and notes. What I found really nice was the guided tour provided at the end of the home page. The tour takes you from one area and explains it more thoroughly. The tour has its own links which I feel provides a lot of information. In order to get the most from this site, the reader must probe deeper into the links. Each link has more new and exciting information that one cannot see from the home page.

Title: Where the Waves Are
URL: http://www.discovery.com/news/features/surfing/surfing.html
Grade Level: Grades 9–12
Best Search Engine: www.yahoo.com
Key Search Word: waves
Review: This is a great interactive site for students to learn from. There is a place to create your own wave with a virtual "game." Here you can choose the variables to learn what creates the biggest wave or the longest one. It is a lot of fun. There is also a section on wave prediction. It discusses the practical uses of predicting waves and it teaches how you can do it yourself. The final section deals with the physics of wave riding. It discusses the shape, speed consistency, glassyness, and size of different waves and which of these variables create the best wave to ride. There are related links to other wave pages and great pictures of different waves. This site is not the best to educate on

waves and how they are created, but it is definitely a lot of fun. This site could be used in science classes as an extra activity, not as the main part of the lesson.

Weather

Title: Everything Weather—Weather Science
URL: http://www.everythingweather.com/weatherscience.shtml
Grade Level: High school and adult
Best Search Engines: www.metacrawler.com; www.yahoo.com
Key Search Words: weather, weather science
Review: This Web site was created in 2001 by Weather Edge, listed as webmaster. Its purpose was to provide a reference for people interested in weather. A variety of sites were created to provide general weather information and presentations on weather topics. Although no individual is cited as the source of the Web site information, each online presentation identifies its individual author. Since no more than their name, title, and affiliation is given, one cannot really assess their credibility. Some of the presentations are dated but otherwise no update information for this site is given. Most interesting are the presentations on storm chasing with pictures included. The weather science sections get very technical and include research on Rear Flank Downdraft effects, hail, cloud electrification, lightning and other charges, radar identification, satellite images, surface maps, and weather software. There is a subscription and mailing list available. The Web site domain is neither government nor education affiliated.

Title: Intellicast
URL: http://www.intellicast.com
Grade Level: High school and college
Best Search Engine: www.google.com
Key Search Word: weather
Review: This advanced level Web site offers an enormous amount of information on the topic of weather. It contains over 250,000 pages of current information on this topic and is visited by millions of people monthly. It has areas containing current weather maps, latest weather headlines (both national and international), and weather planners with regards to travel or sports. The most useful section I discovered after moving through this site was the area titled Education. This area is wonderful for the advanced learner. It contains an Article Library for research; Weather Q&A, which answers numerous technical

questions about weather; Weather 101, which contains an extensive glossary of weather terms (many of which I am unfamiliar with); Seasonal Currents, which has many articles listed thematically by season; and Climate Watch, which contains a wealth of information, much of which is based on studies done by NASA. It contains several oceanic maps, articles on El Niño, and so forth. When I first began searching this site I felt it was great for the average weather-conscious adult; after stumbling on the Education area, I now think it would be great for high school level students or even college level students studying meteorology. I strongly recommend it.

Title: Yahoo Weather
URL: http://wwwa.accuweather.com/adcbin/public/index.asp?partner=accuweather
Grade Level: Middle and high school
Best Search Engine: www.yahoo.com
Key Search Word: weather
Review: Great site with a wealth of information regarding the weather. In one area the user can select a feature. Those topics include Hurricane center, Maps, Radar, Satellite, Storms, News and features, World, What's new, Desk top alert, Site weather cams, Mobile, and Stream. The site has nice graphics and maps. There is also a Today's weather headlines. I was upset when I went into this link predicting this winter's weather and they wanted me to pay for the service. However, other spots under Weather headlines were fine. There is also local weather information involving travel, ski, health, marine, sports, garden, meteorologist, aviation, agriculture, world precipitation, planner, maps, radar, news, and features. All of these sites have nice graphics. Besides having to pay for the winter forecast, there were too many advertisement links at this site.

Title: Weather Classroom
URL: http://www.weather.com/education/wxclass/wxinsights.html
Grade Level: Elementary through high school
Search Engine: www.yahoo.com
Key Search Word: weather
Review: This site is sponsored by the Weather Channel and is developed for students and their teachers. This in an interactive site that not only educates about the weather, but also lets students ask questions to the weather people at the Weather Channel. There are a variety of classroom suggestions for teachers with follow-up suggestions and links to

more information. Some of the activities include developing a weather and people concepts map, an activity called "I Spy Something Affected by the Weather," and The Weather and Me. These activities teach children how their lives are affected by weather. On the School day page, teachers can access weather maps for their area and students can post weather questions. There is also a weather question of the week, in addition to photos, and more weather information. This is a great site that provides a wealth of information for teachers and students.

Title: Weather Education and Safety Homepage
URL: http://www.erh.noaa.gov/er/lwx/wesh/
Grade Level: K–5
Best Search Engine: www.google.com
Key Search Words: weather education
Review: This is an excellent Web site. It had a great deal of information, presented in a way students will understand. The site was actually divided into a Kid section and a teacher section. Under Just for Kids, there were categories such as weather terms and definitions, aviation weather, daily weather quiz, tornadoes, and thunderstorms and lightning. If you clicked on Just for Kids, there were at least 20 weather-related topics to click on. The teacher section had categories such as, weather forecasting and analyses, U.S. interactive climate page, weather calculator, hurricanes, and the hydrologic cycle. This site was so comprehensive, I think it would be a wonderful resource for all weather inquiries.

Title: Wind
URL: http://www.wxdude.com
Grade Level: Grade 3 and up
Best Search Engine: www.ixquick.com
Key Search Word: weather
Review: The main Web site features Nick Walker: "Weather Dude." The main page also indicates that he is on the Weather Channel and is a nationally known meteorologist. This Web site is a very informative, interactive, and educational Web site for kids, parents, and teachers. There is posted news that says "Attention Teachers: a $500 weather grant is now available, get an autographed copy of Nick's sing along with the Weather Dude." The Web site also has an interactive icon that can indicate the forecast for your area just by entering a city or zip code. This seems to be a user-friendly Web site. It features many icons to click on. It also has songs to listen to and purchase. Further-

more, this Web site has won many awards and is a Net Mom–approved Web site. I recommend this Web site to everybody. Other users will enjoy the songs like I did!

Weathering

Title: University of Hawaii Weather
URL: http://www.hawaii.edu/news/localweather
Grade Level: All grade levels
Best Search Engine: www.google.com
Key Search Word: weather
Review: Just because it is December and I am stuck in New Jersey, my site for weather is none other than the University of Hawaii. To make things a little more interesting for me, this site gives local surf reports for the area. I wonder how much teachers make in Hawaii. This site is not fancy or the most creative. It is blunt and to the point, but who doesn't want to know what the surf is like on the Island of Oahu. There is, however, a section for satellite imagery for those who are actually going to use this site for research. It is also helpful for the lucky ones who are able to go on vacation to Hawaii. My favorite part of this site is the five-day forecast, where it is going to be hot and sunny for the rest of the week. It looks like they wrote a forecast for one day and then cut and pasted it over and over for every day.

Title: Weathering
URL: http://www.yahoo.com/bin/search?p=weathering
Grade Level: Grades 11–12
Best Search Engine: Surfwax.com
Key Search Word: weathering
Review: This is written for a higher level, grades 11 or 12. There is a link to the university that sponsors the Web page as well as to the author. The site is four pages long with an in-depth overview of weathering. Their are enough terms and examples to break a class into co-operative groups and allow different areas of research to tie into the topic of weathering. Three main areas of weathering are covered: Chemical Weathering, Biological Weathering, and Physical Weathering.

Title: Weathering
URLs: http://www.geography4kids.com/files/land_weathering.html and http://pasadena.wr.usgs.gov/office/ganderson/es10/lectures/lecture11/lecture11.html

Grade Level: Elementary and middle school
Best Search Engine: www.google.com
Key Search Word: weathering
Review: The first Web site, Geography4kids, features definitions for Mechanical, Chemical, and Biological weathering. Each area features a detailed description of the particular type of weathering and its causes and effects. There is a picture for each type of weathering. The language is very easy to understand. This site can be used as an introduction to the topic of weathering. I included this second site because the first site was very brief and limited. The second Web site is basically lecture notes about weathering. A definition of weathering is provided, along with the types of weathering. Chemical weathering offers descriptions of the three types of weathering: Dissolution/Leaching, Oxidation/Rusting, and Hydration. Biological weathering, or weathering done by living things, gives examples of how tree roots, bacteria and chitins/limpets can cause weathering. The site also describes weathering rates and ways in which the types of weathering help each other out. There are no graphics on this site.

Title: Weathering
URL: http://www.gpc.peachnet.edu/~pgore/geology/geo101/weather.htm
Grade Level: AP high school and college
Best Search Engine: www.metacrawler.com
Key Search Word: weathering
Description: This site was created by Pamela Gore of Georgia Perimeter College. First, the site explains the difference between weathering and erosion. Then it contrasts chemical and mechanical weathering. Finally, the factors that influence the type and rate of weathering are listed and discussed. The site is clearly broken down using an outline format. Information is given through brief items underneath each major topic heading. Excellent pictures help the user understand weathering. There is a lot of information and users need to have some background knowledge on weathering or this site would prove to be quite difficult to comprehend. The site also links the user to other sites on the topic on geology.

Title: Fundamentals of Physical Geography
URL: http://www.geog.ouc.bc.ca/physgeog/contents/11b.html
Grade Level: AP high school, college freshman
Best Search Engine: www.metacrawler.com
Key Search Word: weathering

Review: This Web site was created by Michael J. Pidwirny, Ph.D. of the
Department of Geography, Okanagan University College. He provides an e-mail address for suggestion and corrections. The opening
page begins with the title beneath which is a toolbar divided into
contents, glossary, study guide, links, search, and instructors. Beneath this is an article on weathering. It contains a definition of
weathering, an explanation of the products of weathering and the
three broad categories of mechanisms for weathering; chemical,
physical, and biological. Within the article are bold blue words that
when clicked on take you to a glossary. If you go back to the toolbar
and click on Contents you are taken to a table of contents with
chapters related to the study of physical geography. Weathering is
actually a topic within chapter 11: Introduction to Geomorphology.
The glossary contains an extensive list of terms related to physical
geography. There are study guides organized according to the chapters, and there are related Internet links also organized according to
the chapters. Clicking on Search takes you to a site that explains
how to search the Web. Clicking on Instructors takes you to a page
specifically for instructors of introductory university courses dealing
with physical geography. This Web site is actually Professor Pid-
wirney's online textbook. It was easy to navigate and I found it easy
to understand. The topic of weathering, however, is only one topic
within one chapter and it contains introductory material. This site
could be useful to someone teaching a course on physical geography,
but if your interest is in-depth study on weathering this isn't the
best site. The Internet links do not specifically address the topic of
weathering so for in depth information I would probably bypass
this site.

Title: Weathering
URL: www.geography4kids.com
Grade Level: Middle school and above
Best Search Engine: www.google.com
Key Search Word: weathering
Review: This site is designed for teachers and older children. It is a site
that has many areas of interest in science for children to study and
learn from. The opening page starts out with a section on mechanical weathering, followed by chemical weathering, and then biological weathering. The site includes the following links: biosphere,
global geology, ecology, ecosystems, biomes, populations, food
chains, cycles, soil, and erosion. The following example links are

provided: dune formation, frost action, hillside erosion, and land-slides. More information is provided on B4K microbes, and B4K lichen. The site goes on further with a tour that takes you to other areas of interest; the first stop on this tour is the terrasphere. The tour continues on with other information that is important to the topic of weathering.

Title: Weathering
URL: http://www.peachnet.com
Grade Level: High school and above
Best Search Engine: www.metaspider.com
Key Search Word: weathering
Review: The site is packed with information on the first page. It does not have many links. The few that it has are located at the end of the report. The page starts out with site objectives. Then it moves directly into The Rock Cycle with the explanation of cycle. The site next moves to the definition of weathering and talks about the types of weathering. The following headlines are addressed in this site. Each area shows really nice and detailed pictures. The topics are Physical and Mechanical Weathering, Chemical Weathering, and Biological Weathering. The site further goes on to tell about what happens when granite is weathered. In addition to the topic headings is the Goldich Stability Series. The following links are provided: Physical Geology Online and Physical Geology, GSAMs Page, and the Georgia Geo Science Online.

Title: Weathering
URL: http://www.gpc.peachnet.edu/~pgore/geology/ge
Grade Level: Grades 4–12
Best Search Engine: www.aol.com
Key Search Word: weathering
Review: This site is a good introduction to rocks and weathering. It begins with a discussion on the types of rocks and the rock cycle. Then it goes into more detail about weathering. It defines weathering and it differentiates between biological, mechanical, and chemical weathering. There is also a discussion on the difference between weathering and erosion, and finally ends with the products and factors of weathering. There are some great pictures so students can see different types of rocks and the product of weathering on them. This is a great site to get a lot of information from.

Wetlands

Title: Information on Wetlands
URL: http://h2osparc.wq.ncsu.edu/info/wetlands
Grade Level: Middle school to college
Best Search Engine: www.aol.com
Key Search Word: wetlands
Review: This Web site contains information on wetlands provided by North Carolina State University. This site is an in-depth explanation of wetlands and all information that goes along with wetlands. I focused particularly on Importance of Wetlands: Functions and Values. As you click onto the site you come across an outline with nine sections. Each part of the outline gives different information on wetlands. The site opens up with a menu page divided into nine major topics: Wetlands Introduction; What Are Wetlands?; Definitions and Classifications; Importance of Wetlands: Functions and Values; Importance by Wetland Types: Watershed Roles; Human Impacts: Wetland Loss and Degradation; Wetland Protection: Government Programs; Regulatory Last Resort: Mitigation; Wetland Management: For the Preservation of an Ecosystem; and Links to Other Wetland Information Sources. The function is the process while the value is its meaningfulness to the environment. This section described, in depth, the functions and values of the wetlands. It also had a table comparing the two. It is easier to understand the difference between a function and a value when they are laid out in front of you in an organized fashion. I was interested in the section on Wetlands Loss and Degradation. The causes of the loss include conversion to farmland, residential development and road construction. These are just a few, but wetlands can also become impaired. There is a subsection on Wetland Impairment. This was a very informative Web site. If I were a science teacher, I would use this site. Each topic provides a great deal of information on its subject and provides additional links to more in-depth information on the topic. Navigating this site is easy, and you can return to the menu page for the bottom of any link you go to. However, there are no bells and whistles here. It is straightforward information. It's not easy on the eyes but it's a good resource.

Title: Dynamics of Wetland
URL: http://www.nsen.unl.edu/lesson/water/teachwetland/html
Best Search Engine: www.profusion.com

Grade Level: Grades 9–12

Key Search Word: wetlands

Review: This site is maintained by The School of Natural Resource Science University of Nebraska–Lincoln. There is a link provided to the author of the site. This site offers a brief description of wetlands, however it offers excellent supplemental aids to lessons. It includes links and projects. It provides activity enhancement including additional research, adopting a wetland, visiting wetland sites with objectives, and student-based monitoring of changes in two different wetlands. I like this site to further develop a student's or class's curiosity about wetland studies.

Title: Wetlands

URL: http://www.epa.gov/owow/wetlands

Grade Level: High school and college

Best Search Engine: www.google.com

Key Search Word: wetlands

Review: This site from the Environmental Protection Agency features three main sections; What Are Wetlands, Why Protect Wetlands, How Are Wetlands Protected and What You Can Do to Protect Our Vital Resource. The What Are Wetlands page offers a definition of what wetland is, along with links to the various types of wetlands, such as marshes, swamps, bogs, and fens. The links provide a picture of the wetland and/or the animal life that inhabits it, a description of the wetland, its function and values, and its status. Why Protect Wetlands features several links that offer information about the function and value of various wetlands, the types of fish and wildlife of wetlands, wetland erosion, water quality and hydrology. How Are Wetlands Protected also features several links that provide information about watershed planning, water quality standards, and restoration. Web site visitors are given many suggestions on how to help protect the wetlands. This link offers suggestions such as buying federal duck stamps from your post office to support wetland acquisition, participating in the Clean Water Act Section 404 program and state regulatory programs, or Adopting a Wetland. There is also an In the News section, which features recent articles relevant to wetlands and their preservation and restoration, and several links to pages that detail the current laws and regulations regarding wetlands. This is an excellent site. The links are plentiful and are a great resource for anyone interested in learning about wetlands or helping to reserve them.

Title: Wetland Kids
URL: http://www.wetland.org/kids/Kids.htm
Grade Level: Grades 4–7
Best Search Engine: www.metacrawler.com
Key Search Word: wetlands
Review: Wetland Kids is a site created by Environmental Concern, Inc. It is
   a nonprofit organization dedicated to wetlands restoration, research,
   and education since 1972. The home page is colorful and eye-catching.
   The user can click on several areas to learn about wetlands. The following links are available: What is a wetland?, What wetlands do for
   you!, Activities, What you can do!, Find a wetland near you, Did you
   know?, Wetland words, and More links. The Activities link has a message that says Coming Soon. Each other link on this site has a ton of information written in a clear manner. There are many pictures
   incorporated onto every page. The site gives users tips for how to save
   the wetlands and lots of facts about wetlands. The Wetlands Words
   section is excellent! There is a comprehensive list of words related to
   wetlands that will be a great benefit to teachers and students. I recommend this Web site to any teacher/student studying a unit on wetlands.

Title: Wetlands
URL: http://www.epa.gov.
Grade Level: Grade 8 and up
Best Search Engine: www.google.com
Key Search Word: wetlands
Review: For the person requiring a lot of information on wetlands, this is
   the perfect site. This site is a government site. The information is current as well as accurate. The page opens with topics that have links to
   What are wetlands?; definitions, types, status and trends Why protect
   wetlands?; functions and values, fish wildlife, flood protection, shoreline, erosion, water quality hydrology and economics; How are wetlands protected? watershed planning, water quality standards,
   migration, financial assistance, and restoration. The site contains a
   drop-down menu that has links to 38 topics relating to wetlands. In
   addition there are two sidebar links that contain laws, regulations,
   guidance and scientific documents. State tribunals and initiatives.
   Landowners, water quality, and 401 certification monitoring and assess wetlands and watersheds, restoration, education, information in
   your area. All of this is included on one sidebar link. The other sidebar
   link contains In the news, awards and programs, fact sheets, photos

and calendar of events. I think this site is loaded with information and will be quite useful for someone wanting to learn a lot on the wetlands.

Winds

Title: Wind
URL: http://www.awea.org/default.htm
Grade Level: Grade 3 and up
Best Search Engine: www.ixquick.com
Key Search Word: wind
Review: The main Web site features the headline: American Wind Energy Association. There are many pages to enter from the main page. On the left side of the main page is About AWEA, News and releases, What's new, and Small wind systems. On the right side of the page are Utility-scale wind, AWEA member's directory, Energy policy, Resource library, and Green power. There are more links underneath each right and left side. The center of the page features many Web sites about the American Wind Association. These sites are very informative for children in third grade and up. This site has bountiful information on wind. It may seem overwhelming for the first-time user. My recommendation is to always use a left mouse click to move back. This Web site also features things to purchase as well as how to become a member. I found this Web site to be helpful with educational reports and projects. I also discovered the coloring book is fun to color and seemingly accurate. The authors of this educational Web site are from the American Wind Association. All the links for kids are age appropriate for children in grades three and up. I found the main Web site to be helpful and interactive for children in grades 3 and up. This is also a beneficial Web site for teachers.

Title: Wind
URL: http://sln.fi.edu/tfi/units/energy/wind.html
Grade Level: High school
Best Search Engine: www.google.com
Key Search Word: wind
Review: This site is about the science of wind technology. It opens, "A gentle breeze cools our home on a hot summer day. . . . With sails unfurled, a sailboat races toward the horizon. The wind is its friend." It continues: "Surf's up and the waves pound wickedly, whipping the sailboats toward the rocks. The wind is fierce." With the exploration of the science of wind energy we can learn more about our relationship

with wind. The section titled Blistery Beginnings allows for a photo gallery, video gallery, online resources, "windy booklist (stories of the greatest wind storms), figurative language, and windy things to do. This section expands the spectrum of wind energy. Studying wind can be fun and not so academic. Other sections involve the research behind wind energy. Early exploration of how wind effects us. Inquiry-based learning can produce a richer understanding of wind energy.

Title: Forces and Winds Online Meteorology Guide
URL: http://ww2010.atmos.uiuc.edu/(Gh)/guides/mtr/fw/home.rxml
Grade Level: College
Best Search Engine: www.ask.com
Key Search Words: learning about the wind
Review: This Web site was a comprehensive informative site for college students interested in learning about the wind. The site began with a brief introduction. It was then divided into sections: Pressure, Pressure Gradient Force, Coriolis Force, Geostrophic Wind, Gradient Wind, Friction, Boundary Layer Wind, Sea Breezes, Land Breezes, and Acknowledgments. Each of the wind sections had in-depth explanations, charts, diagrams, and tables. The information was presented in a way that young students would have a difficult time comprehending. I could see the benefit of this Web site for a college meteorology course.

Title: Wind Resource Database
URL: http://www.nrel.gov/wind/database.html
Grade Level: High school and adult
Best Search Engine: metacrawler
Key Search Words: wind, wind energy
Review: This Web site was developed out of the U.S. Department of Energy for Wind Technology. This particular site describes how wind is used for energy and how the NREL (National Renewable Energy Lab) identifies and gathers data for wind resource maps of the United States and foreign countries. These maps help users find areas worthy of wind resource monitoring. The NREL has, in addition to maps, developed the Wind Energy Resource Atlas of the United States and its territories. If one has access to Acrobat Reader, there is a paper available through this site that describes how much electricity could be generated from wind in the United States. The National Wind Technology Center is located in Golden, Colorado. It is connected with the following organizations: Meteorological Tower Information Cen-

ter, the National Center for Atmospheric Research, the Weather Channel, and *USA Today* Weather. There are links to each of these sites included. The section on Wind Energy in the U.S. Marketplace addresses how these labs are helping to bring wind energy, a clean low-cost generation option, to market in the United States and abroad. The American Wind Energy Association is the trade organization for the wind energy industry. It offers a publications catalog, a FAQ section, and articles in a newsletter. This Web site is well organized and well researched, and it provides engaging color pictures and color-coded text. I found it fascinating because of its link to The Weather Channel, which I rely on daily.

Title: Wind
URL: http://www.fi.edu/tfi/units/energy/blustery.html
Grade Level: Elementary and middle school
Search Engine: www.yahoo.com
Key Search Words: wind & earth science
Review: This is a great site that is sponsored by Franklin Institute. This is a site for teachers that offers suggestions for lessons about the wind. Several lesson plans are provided including "What's the Wind" and "Building the Best Windmill." These lesson plans are complete. Each offers the following sections: 1. Prior knowledge, 2. Exploration, 3. Explanation, 4. Examples from everyday life, and 5. Assessment. This site provides information about motivational resources for teacher. It also makes suggestions for related activities and allows students work to be posted on the site for others to access. Teachers can post their students' stories and pictures of projects, and the site will allow posting of suggested design projects such as a wind-resistant building and wind-powered vehicles. This is a very neat site for teachers.

Title: Whirling Winds of the World
URL: http://freespace.virgin.net
Grade Level: High school
Best Search Engine: www.google.com
Key Search Word: winds
Review: What a neat Web site!! I am pleased to have finally stumbled upon an appropriate "winds" site, as many were about musical instruments. Anyway, this site, created by webmaster Mike Ryding, is so interesting. (He can be e-mailed with comments or corrections.) It contains four areas: Global, Seasonal, Local, and Spinning. The

Global section covers winds types such as jet streams, trade winds, westerlies, and polar easterlies, and it offers great descriptions of each. The Seasonal section is divided into areas such as monsoons (annual winds) and daily winds such as anabatic, katabatic, valley winds, and sea breezes. The Local section contains a glossary of hundreds of terms related to the topic of winds. Lastly, Spinning covers topics such as cyclones, sand devils, dust devils, tropical storms, twisters, typhoons, and so forth. The list is very lengthy and well described. This site overall is very comprehensive and would serve a high school student well.

3
SUPPLIES

Whether you are looking for a microscope to use in your research, or a specific slide necessary for you to help your child do a science activity at home, earth science supplies and information on the best places to order supplies are a must. This section will focus on earth science supplies.

Title: Carolina Biological
URL: http://www.carolina.com
Grade Level: All grade levels
Review: Carolina Biological has an online catalog that allows you to browse at your leisure. Scroll down the left menu to "biology" and click on the link that will take you to the biology section of the catalog. The menu also has supplies for the K–6 science teacher and the middle/junior high science teacher. Each item is accompanied with a picture. A click on the item provides more information about the item, including the cost and a link to order. This is a user-friendly way to order supplies from a reputable business.

Title: Science Stuff
URL: http://www.sciencestuff.com
Grade Level: Grade and middle school
Review: Four links on the left side of the home page are "chemistry," "toys," "environment," and "laboratory." When you click on each link, new subtitles appear. For instance, clicking on "chemistry" provides access to "models and charts." The home page also has a sales link to items that are reduced in price. This supply house is limited but worth a look for the sales link.

Title: Fisher Science Education
URL: http://www.fisheredu.com
Grade Level: All grade levels
Review: Fisher Science Education is more than a supply catalog. At the
    home page, the site highlights science news from worldwide news
    sources including CNN and the *New York Times*, a science fact of the
    day, and "from my classroom to yours." In addition there are products
    on sale and products announcing highlights. At the top of the home
    page are links to the catalog, teacher resources, and Fisher worldwide,
    to name three of the many links. When you place your mouse arrow
    on each of the links, a sublist appears. Teacher resources have teacher
    tips, science calendar, science-ed directory, and FAQs. The online
    catalog is extremely easy to use. There is a browse button and a search
    window in which you can type your needs either specifically or gener-
    ally. For instance, if you type the word "biology" you will be provided
    with a long list of items that are biology related. The same shorter list
    will appear if you type the more specific term "thermometer." Click-
    ing on one of the items in the list takes you to the specific page of the
    item with a picture and other information. You can order by clicking
    on the "add to my cart" button. The science calendar link is worth a
    look. It is organized by month and gives you science information,
    such as science inventions and scientists' birthdates, for each day.
    Clicking on "teacher tips" leads you to a menu listed by discipline,
    such as biology or chemistry. The tips contain demonstration and lab-
    oratory activities. The science-ed directory provides the address and
    phone numbers of science-related programs, museums, and agencies
    listed by state. This Web site is extremely well maintained and will
    provide more than supply information.

Title: Chemical Online
URL: http://www.chemicalonline.com
Grade Level: College
Review: This Web site provides good reviews of the chemistry books the
    company sells. Clicking on "Buy on Line" leads you to a number of
    links. The most interesting is the link to LabEx.com, which sells used
    laboratory equipment. Back at the home page, clicking on the "news
    and community" link leads you to free software to download, discus-
    sion forums, and an events calendar. Even though this site focuses on
    industrial chemistry, the books, and news and community are worth a
    look.

Title: Sargentwelch.com
URL: http://www.sargentwelch.com
Grade Level: High school and above
Review: Sargent Welch and VWR can be found on this Web site. This is
the site to use to obtain MSDS (Material Safety Data Sheet) information concerning specific chemicals, which is required by OSHA (Occupational Safety and Hazard Administration). The home page
provides links to the complete catalog, safety, grants, correlations,
Web partners, MSDS, and science references. The correlations link
provides lists of supplies needed for various curriculums. For instance,
there is a high school planning list containing the name of the material, the catalog number, and price per unit. This spreadsheet list
would come in handy when ordering year to year. There are correlations for high school, middle school, AP biology, ChemCom, AP Environmental listing, Biology: A Community Context Materials List,
EZ Prep, Science Fair, and Harcourt Brace elementary science books
for grades 1–5. The Sci references are a great resource linking you to
general resources, resources by discipline and subject, program and organizational resources, general science educational resources, and
newsgroups related to science education. The Web partners lead you
directly to partners in various specific areas such as Science Links and
Ohaus microscopes, to mention a few.

Title: Science Kit and Boreal Labs
URL: http://www.sciencekit.com
Grade Level: Middle school through high school
Review: This company claims to be "the leading supplier of science materials and equipment to science teachers throughout the United
States." At the home page, users find links to classroom activities
provided in all disciplines, a recommended materials list link that allows you to type in the title of the textbook that you are using in
order to receive a list of materials that are recommended for use in
the activities, and a markbook download containing seating plans
and other organizational tools. You can order a catalog and place an
order through the order assistant; however, there is not a catalog
available online, which is a definite disadvantage.

Title: Ward's Natural Science Establishment
URL: http://www.wardsci.com
Grade Level: Grades 4–12

Review: Ward specializes in biology, geology, and life/environmental earth/physics subjects. Ward's catalog is not online, but a complete list of subtopics is available by clicking on the subjects at the home page. You can, however, view the company's new products by clicking on the new product link and then on the specific product.

Title: Daigger Laboratory Equipment and Supplies
URL: http://www.daigger.com
Grade Level: High school and beyond
Review: Daigger's catalog has to be the easiest and most user-friendly supply house catalog. You can search the catalog by clicking on the alphabetical keys, or by typing in the name or catalog number in a search window, or finally by browsing the catalog virtually. The company has 75,000 products in its catalog. If you click on "c" for chemicals, you will be led to a list of science materials all beginning with the letter "c" starting with calculators and ending with chemicals (after cylinders). If you know the name of the chemical you are searching for you can click on the first letter of the chemical name or scroll down an alphabetical list of chemicals until you reach your product. Clicking on the product provides you information and ordering prices, as well as MSDS sheets.

Title: Edmund Scientific Company
URL: http://www.scientificsonline.com
Grade Level: All grade levels
Review: Edmund's Scientific to a scientist is like a candy store to a candy lover. There is the usual catalog by topic link list on the left of the home page. Unique to this site is the gift search link, which allows you to search for a science gift by topic or by price. Great idea! You can buy a drinking bird for only $7.95. For those of you looking for "deals," there is a clearance link. The tech tips is invaluable for hints on using technology equipment. Topics like "how to read a binoculars power" provides the novice with expert advice and information.

Title: Educational Innovations, Inc.
URL: http://www.teachersource.com
Grade Level: All grade levels
Review: "The Master Teacher Source for Science Workshop Supplies" is the goal of Educational Innovations, Inc. The company was started to provide science workshop professionals with one-source shopping for their science supplies. At the home page, find links to the extensive

catalog and to a schedule of shows and conferences where the company will be holding workshops. Their catalog is divided into subjects and topics. A click on the life science/biology link leads you to a more specific subtopic index. Because this site was designed for people who do workshops, the prices for the items decrease with the increase of quantity. Usually you order in quantities of 1–10, 11–29, and more than 30.

Title: Flinn Scientific
URL: http://www.flinnsci.com
Grade Level: High school
Review: This great resource site includes links to chemistry, biology, and safety with proven solutions to problems, Flinn freebies, and laboratory design for designing a current laboratory, computer interface technology, and Flinn scientific order maker. I did not find Flinn's catalog online. I think that you have to download it after clicking on the Order maker link. This is a definite problem with this site. However, all the other links provide wonderful resources. The Safety link even has an e-mail conference for use with your science department at departmental meetings. The information provided with the laboratory design link is wonderful and could be useful in order to upgrade a laboratory. Flinn freebies look to be wonderful resources but are only for certified high school science teachers.

Title: Pasco Scientific
URL: http://www.pasco.com
Grade Level: Middle school to high school
Review: Pasco Scientific has the goal to design, manufacture, and service the highest-quality products for science teachers worldwide. Clicking on "Products" leads you to the online catalog, which is indexed according to subject and topic. You can browse the complete index or type in a topic at the search window. Unique to Pasco is the Experiment Central link. By clicking on either the Data Studio Library or the Science Workshop Library you can download over 400 experiments. A click on Biology Experiments grades 9–12 will allow you to download 33 biology experiments. In order to do so you will need compression software to unstuff the files.

4
MUSEUMS, SCIENCE CENTERS, AND SUMMER PROGRAMS

Science can be found in general museums and science centers or in specific organizations like nature centers and parks, zoos and aquariums, and planetariums and observatories. This chapter will provide you with links to Web sites that provide directories of these resources with direct links to the site. In addition, the best sites will be reviewed. Science summer programs are usually either locally or nationally supported. Science museums/centers no longer are drab storage areas of objects with "do not touch" signs. They are "fun" resources for informal learning. They are places to touch, discover, and explore. There are more than 400 science centers and museums around the world. These can be found at the Association of Science and Technology Centers Web site at http://www.astc.org. A review of this site will be provided later in this chapter. In addition, a directory of all-formal science museums/centers, nature centers/parks, zoos/aquariums/aviaries, and planetariums/observatories can be found at Science Adventures Web site at http://www.science adventures.org.

Title: Association of Science and Technology Centers
URL: http://www.astc.org
Grade Level: All grade levels
Review: This is a Web site for science museum professionals. It provides
 both information about the ASTC organization and also resources for
 developing exhibits and resource materials for the science museum
 worker. Even if this part of the site is focused on the museum professional, there also is a link to "find a science center" that provides you
 with a search page where you can search for a science center first by

country then by state. The link will then give you a direct link to the science center plus its address, phone, and fax numbers. If you don't want to find a science center but wish to explore, you click on the "click here to explore" link, click on "try science," "parent," or "site map" links or type in a subject in the search window. The "try science" link leads to live interactive sites that allow you to explore specific topics live. For instance, you can watch the penguins at Montreal Biodome or observe the panoramic view from New Hampshire's Mount Washington. Very cool!

Title: Earthcam
URL: http://www.earthcam.com
Grade Level: All grade levels
Review: This is the reference to use to search for real-time science and exploration. The main menu gives you a number of choices, including Education Cams, Weather Cams, Space and Science Cams. When you click on Weather Cam, you are able to access 797 real-time cam sites. You can narrow that number at the search window by typing in your interest/topic or clicking on the 10 subject links: forecast, outdoor, ski and surf reports, kids, streaming, weather, indoor, remote control, news, and seismograph. The 797 referenced Web sites contain a linking URL and a description of the site. For example: the 150-foot solar tower cam gives you a view from Mount Wilson, California. The image is captured every four minutes from the Mount Wilson Observatory. It is a spectacular view! The link provides accurate weather conditions; for instance, the weather during this visit is recorded at clear, calm; 78° F. The Web cam has links for you to learn more about the lab's research; go to the observatory homepage at UCLA's Department of Astronomy.

Title: Science Adventures
URL: http://www.scienceadventures.org
Grade Level: All grade levels
Review: This site has a vast reference base for about 1,800 links to museums/science centers, nature centers/parks and gardens, zoos, aquariums and aviaries, and planetariums and observatories. It was developed by the Eisenhower Regional Consortia for Mathematics and Science Education. Clicking on any of the above four topics or typing in your own topic at the search window leads you to a list of Web sites alphabetized by state. Clicking on the title of the organization, for example, "California Science Center," provides addi-

tional information, including a direct link into the center's Web page, a description of the center, contact information including phone number, and other information including reservation availability and procedures.

Title: Exploratorium
URL: http://www.exploratorium.edu
Grade Level: All grade levels
Review: This museum was founded in 1969 by noted physicist and educator Dr. Frank Oppenheimer. The mission of the museum is to "create a culture of learning through innovative environments, programs, and tools that help people to nurture their curiosity about the world around them. It is housed within San Francisco's Palace of Fine Arts. The home page of the Exploratorium offers myriad live Webcams, including watching a total solar eclipse from Zambia! You can even visit the movable live roof cam or exhibit cam. The Exploratorium is an interesting site to visit to view and understand the latest innovative technology. Its extensive exploration into live Webcams makes this museum unique and worthy of visiting online or in person.

Title: Educational Outreach at JPL
URL: http://eis.jpl.nasa.gov/eao/students.html
Grade Level: Middle school through high school
Review: This program site of NASA at the Jet Propulsion Laboratory in California provides a list of programs for kids mostly provided via the Internet. Cassini's Kids' Corner is written by kids for kids. Information about the Cassini spacecraft and Saturn includes both easy and advanced model plans of the spacecraft for you to build. Other links include KidSat, the Space Place, and TOPEX/Poseidon Educational Outreach.

Title: Wonderama "Explorations in Science Education"
URL: http://www.wonderama.org
Grade Level: Ages 5–14
Review: Wonderama is an East Lansing, Michigan, corporation that provides hands-on summer science programs at low cost to local Community Education Centers. Even though the programs are limited to the Michigan area, you can use the Web site, which provides experiments, gidgets and gadgets (cheap materials), and links to Student Works.

Title: SHARP: Summer High School Apprenticeship Research Program
URL: http://mtsibase.com/sharp
Grade Level: High school (at least 16 years old)
Review: Every year NASA operates an eight-week summer program for
   high school students at their field installations. Students are selected
   for their aptitude and interest in science and engineering careers.
   Eligibility in SHARP includes being at least 16 years old by the start
   of the program in June, a U.S. citizen, completion of two college
   preparatory mathematics and science courses, strong aptitude and interest in a career in science, engineering, or technology, permanent
   resident of the state of the field installation, willingness to participate
   in a formal interview, and availability full-time Monday through Fri-
   day 40 hours weekly throughout the entire eight weeks. Field installations are in California, Texas, Alabama, Ohio, West Virginia, Florida,
   Maryland, and Virginia. NASA also has a program SHARP Plus
   (QEM) Quality Education for Minorities Network, at http://qemnet
   work.qem.org/sharpplus.html. This program's goal is to increase the
   success and participation of minorities in science and is held on college campuses. Eligibility is based on aptitude and interest in a career
   in science.

Title: Project Seed
URL: http://www.science.sjsu.edu/projseed.html
Grade Level: High school
Review: This project, supported by the American Chemical Society, is
   designed to encourage economically disadvantaged high school stu-
   dents to pursue careers in chemistry. Students apply and those who
   are chosen work in research at a participating college. Application
   forms and specifics about eligibility are available at this URL. This
   Web page can also be reached by accessing the American Chemical
   Society's home page at http://www.acs.org and then typing "Project
   Seed" in the search window. This path seemed easier than typing in
   the long URL for Project Seed.

Title: Boston Museum of Science
URL: http://www.mos.org
Grade Level: All grade levels
Review: This museum has the typical "what's happening," "general information," "store," and "support the museum" links. However, it also
   has an online virtual exhibit link that is worth visiting. On this visit
   the exhibits are the Virtual Fish-Tank, Secrets of Aging, Secrets of

the Ice, Messages, Everest, Leonardo, Oceans Alive, Scanning Electron Microscope, Theatre of Electricity, WeatherNet, Dance of Chance, Science Learning Network, and Big Dig Archeology. Most of these are non-interactive but provide great resources on specific topics. Some are local. The WeatherNet link provides current local weather information including an interactive map.

Title: Science Learning Network
URL: http://www.sln.org
Grade Level: All grade levels
Review: This is an online network of people interested in demonstrating a new model of inquiry education. It was founded by the Exploratorium, Boston Museum of Science, Franklin Institute of Philadelphia, Miami Museum of Science, Science Museum of Minnesota, and Oregon Museum of Science and Industry and each museum's supporting school. It is funded by the National Science Foundation and Unisys Corporation. The group expanded to include international museums, including Heureka, the Finnish Science Center; Science Museum of London; New Metropolis Science & Technology Center Netherlands; Science Museum, Japan Science Foundation; Singapore Science Center; and Exploradome in France. You can directly link to the museums. Each museum has developed resources that are shared online. To access, click on the "explore our resources" link. Many of these resources are available in different languages. The Exploradome in France seems to be only in French.

Title: Franklin Institute Science Museum
URL: http://sln.fi.edu/tfi
Grade Level: Middle school
Review: The Franklin Institute's mission is to promote the public understanding of science. It opened its doors to the public in 1934 in Philadelphia. It includes the Fels Planetarium, the Mandell Center, the Tuttlemann IMAX Theatre, and the Musser Theatre. In addition to the usual links to the programs, store, and resources, the Franklin Institute has a link to "Welcome to my World! Careers in Science and Technology." This link takes you to a menu of career choices including science teacher, geologist, meteorologist, chemist, and computer administrator. For example, you can click on the chemist link and spend a day with a chemist. This is a nice career exhibit for younger middle school students. It also includes links to further explorations in the field of chemistry.

Title: The Field Museum in Chicago
URL: http://www.fieldmuseum.org
Grade Level: All grade levels
Review: The Field Museum's strength is in its research and collections.
    Objects in Anthropology, Botany, Environmental and Conservation,
    Geology, and Zoology are only a few of the major links that you can
    open. When you access any one of these links you are provided with
    the major research that is being done and a brief yet informative de-
    scription of the research. Usually there is another link for more infor-
    mation. At the home page, you can access some wonderful resources
    through the education link. During this visit, programs on Environ-
    mental Rescue, and a virtual underwater expedition are being high-
    lighted.

Title: Smithsonian Museum of Natural History
URL: http://www.mnh.si.edu
Grade Level: All grade levels
Review: The strength of this museum, like the Field Museum in Chicago,
    lies in its research and collections. There are 124 million objects and
    specimens held at the National Museum in Washington, DC. In
    order to search for an object, click on the Research & Collections
    link at the home page. That leads you to a menu on the left; click on
    site map. You can then click on the "databases" link, which takes you
    to Informatics, Collections Databases, Reference Databases, and Bib-
    liographies. The Collections Databases narrows your search to an eas-
    ier searchable list. If you were interested in Fish, you might click
    on the Fish Collection—Online Images. Clicking on an individual
    fish image leads you to a classification and identification key for
    that particular fish. For instance, USNM 00336651 is *Pervagor
    Melanocephalus*, found in the Pacific Ocean off the island of Tonga in
    1993. It has the date collected, by whom, on which vessel, and eco-
    logical information. The research and collection component of the
    National Museum is invaluable for the researcher. The museum also
    offers a number of educational programs, some of which are free. Ex-
    pedition to the Galapagos is a free program that is connected to the
    IMAX film. It includes Photo Journals, Video Footage, and field
    notes.

Title: National Geographic
URL: http://www.nationalgeographic.com
Grade Level: All grade levels

Review: The National Geographic offers the same high-quality material on its Web site that it does in its magazine. There are map, kids, education, forum, and live events links on the home page. Clicking on the education link takes you to a list of resources for teachers. The Online Adventures link leads you to searchable windows in which you type in a subject area, type of resource, and grade level. This research tool will help you find information, videos, and live programs.

5
CAREERS

This chapter concerns itself with careers in science and lifelong learning, and professional organizations are great resources with which you can begin your research. Depending upon your age, educational background, and career interest, you will have a vast amount of Web sites to choose from. Included in this chapter are the best resources for both students in upper high school as well as in college. These Web sites not only focus on professional organizations but also include societies, federal and state agencies, private companies, and nonprofit groups.

Title: Science, Math, and Engineering Career Resources
URL: http://www.phds.org
Grade Level: High school and above
Review: Sounds like this site is only for people with their doctorate degrees, but that is not the case. In fact, the site contains information for would-be scientists from high school on to advanced degreed people. It ranks graduate schools, posts jobs, and allows you to upload your résumé. The recommended links were impressive sites designed by science professionals, most of whom work at professional organizations such as the American Association for the Advancement of Science and the National Academy of Science. This is a new Web site and worth a look.

Title: American Association for the Advancement of Science
URL: http://www.aaas.org
Grade Level: College and beyond

Review: At the main menu, choose "science careers" at the search window. This takes you to the main career Web site, where you can search for jobs, post your résumé, join a job-alert e-mail group, find out about career fairs, read employers' profiles and follow links to their sites, and obtain advice and perspectives. If you click on the advice and perspectives link, you can obtain help with your résumé and cover letter or you can access the past articles on science career advice.

Title: American Society of Limnology and Oceanography
URL: http://aslo.org
Grade Level: High school and above
Review: On the left side of the Web site scroll down to the Student Information link and click on it. This will send you career information including advice on presentations, advice on aquatic science careers, and career links. If you go down to the Educational Resources link and click onto it, you will be given a list of resources. Some are at the college level, but there are a few at the high school level, such as Strategies for pursuing a career in Marine Mammal Science.

Title: Science Career Information for College and High School Students
URL: http://www.geocities.com/capecanaveral/hangar/4707/hs-career.
    html
Grade Level: High school and college
Review: At the main menu are general topics in biology, biotechnology, genetics, health, physiology, and virology. Also included in the main menu are topics in undergraduate teaching, technology transfer, science writing, law, bioinformatics, public policy, science education, and a scientist's guide to traditional and alternative careers. There is a wealth of information for the serious person who is looking for a career move in science or for the high school student to find financial resources and information on science-related careers.

Title: Careers in Science and Engineering
URL: http://www.nap.edu/readingroom/books/careers
Grade Level: High school and college
Review: This Web site is produced by the National Academy of Sciences. The main menu gives you links to notices, staff and guidance groups, acknowledgments, notes on how to use the guide, request for comments, and contents. Clicking on "contents" shifts you to a career guidance index including "what are your career goals," "how can you

meet your career goals," "what survival skills and personal attributes do you need to succeed," "what education do you need to reach your career goals," "how do you get the job that is right for you," "action points," and a bibliography, plus a discussion of scenarios. This Web site at first glance seems to be an excellent career development site for the advanced student. However, at the end of the contents list there is a list of profiles that indeed would benefit the high school student. Profiles include "How does a Geneticist/Molecular Biologist get to be a Patent Lawyer?" and How does a Research Biologist get to be a high school teacher?." These are personal stories that will appeal to all students of science. This site is personal, interesting and informative, and very well written. A must for anyone interested in careers in science.

Title: American Society for Biochemistry and Molecular Biology
   (ASBMB)
URL: http://www.asbmb.org
Grade Level: Grade 12 to adult
Review: "The American Society for Biochemistry and Molecular Biology
   (ASBMB) is a nonprofit scientific and educational organization with
   over 10,000 members. Most members teach and conduct research at
   colleges and universities. Others conduct research in various government laboratories, nonprofit research institutions and industry," the
   site says. The society's purpose is "to advance the science of biochemistry and molecular biology through publication of scientific and educational journals (*Journal of Biological Chemistry, Molecular and
   Cellular Proteomics, Biochemistry and Molecular Biology Education*), organization of scientific meetings, advocacy for funding of basic research and education, support of science education at all levels, and
   promoting the diversity of individuals entering the scientific work-
   force." On the home page, scroll down and click on "education" in
   the left menu. Then scroll to the "career brochure" link to access the
   .pdf file on careers in biology.

Title: Chemsoc.org
URL: http://www.chemsoc.org
Grade Level: College and beyond
Review: This Web site provides a database of career development train-
   ing opportunities in chemistry, plus links and reviews of other sites for
   chemistry careers. It provides information from the United States,
   England, and Canada. The home page is rather confusing, but click

on either "careers and job center," "web links," or "learning resources." At the "learning resources" link click directly on "careers." You then can choose from an A–Z career resource list, which includes items such as "brilliant careers" and "choosing a chemistry degree." At the home page again, click on "careers and job center" to obtain the latest list of jobs, get help with a résumé or application letter, or read about interview techniques. The "web links" link leads to over 3,000 tried, tested and reviewed Web sites. This is an excellent resource for the college and beyond student.

Title: American Chemical Society
URL: http://www.acs.org
Grade Level: High school and above
Review: Type in the word "careers" at the opening search window. This leads you to another list click on "careers and jobs." You now either click on JobSpectrum.org or C&EN Classifieds. JobSpectrum.org, the chemistry careers connection, will post your résumé for a fee. It also has free information on careers in chemistry and about salaries. A click on the C&EN classifieds provides you access to the newsmagazine of the chemical world online. Scroll to the career and employment link on the left of the page and you will have access to the latest articles on careers in chemistry. It's informative and current. For careers specifically in biology, type in "biology careers" in the home page search window and you will be led to 20 links on careers in biology, including C&EN: Employment—Careers in Bioinformatics, VC2: Careers: Environmental Chemistry, VC2: Careers: Biotechnology, to name just a few of the links.

Title: College and Career Preparation tips for NYC High School Students
URL: http://www.fordham.edu/step/dugan/timeline.htm
Grade Level: Grades 9–12
Review: Even though this is specifically for New York City high school students and written by a Fordham University career counselor, it is applicable to all high school students. It is general, not focused on science; however, it includes all the materials any student needs for his or her career development. It includes advice for grades 9 and 10, summer between grades 10 and 11, grade 11, summer between grades 11 and 12, fall of grade 12, and spring of grade 12. It is comprehensive, including standardized test taking, financial aid and scholarship

advice and links. We praise this site highly and recommend it to any high school student in need of career advice.

Title: Usnews.com: Education
URL: http://www.usnews.com/usnews/edu/home.htm
Grade Level: High school
Review: A searchable database reviews more than 1,400 colleges. It includes reports on campuses, financial aid guides, and scholarship information.

Title: CollegeRecruiter.com
URL: http://www.adguide.com/highschool
Grade Level: High school
Review: You can search for a job by choosing a category, location, job level, and type of job. You can post your résumé or go to the career center and ask a question from the experts for free.

Title: O*net on Line
URL: http://online.onetcenter.org
Grade Level: All grade levels
Review: O*net on Line includes a database holding information on skills, abilities, knowledges, work activities, and interests connected to careers and jobs. It is available for more than 950 job occupations according to the Standard Occupational Classification system. This Web site helps the job seeker find jobs that fit their skills, knowledge, and interests; explore different careers; research the steps necessary to secure their dream job, maximize their earning potential, and finally know what it takes to be successful in their chosen field. The site has low-vision and text-only versions. The home page allows you to explore "find occupations," "skills search," "related occupations," "snapshot," "details," and "crosswalk." Snapshot helps you explore a specific occupation through a worker. Details and Crosswalk link you to more information on careers. If you click on Find Occupations, you will be led to a search window. You type in the title of the job. For instance, type in the word "chemist" and click on the Life, Physical Science, and Social Science link and four job titles appear for you to explore. If you clicked on the "Chemists" link, you could find similar job titles or explore the job in-depth to find out what skills and education are needed. This is an invaluable Web site for any student exploring careers.

GLOSSARY

Bookmark
A tool used to store URL addresses for a Web browser's user.

Boolean operations (AND, NOT, OR)
Used to define a search.

Browser software
Most common are Netscape Navigator (Communicator) and Microsoft Internet Explorer (Explorer).

Domain or DNS (domain name system)
The last part of the URL. It provides the name of the organization and the type of organization. An example: In http://www.cs.csubak. edu, cs.csubak.edu is the DNS; cs.csubak is the name of the organization; and edu is the type of organization (educational).

Finger
An Internet tool used to access names of people on other Internet sites.

FTP (File Transfer Protocol)
A common method used to move files between Internet sites.

Gopher
A program that was developed in 1991 and has become second in usage to the World Wide Web.

Home page
The main Web page for a business, organization, or personal site.

HTML (Hypertext Markup Language)

The language used to create hypertext documents. These documents are used on the Web.

HTTP (Hypertext Transfer Protocol)

A protocol used on the Web.

Internet

The worldwide linked computer network system.

Keyword

A word used in a search.

Link

An association between two Web pages or sites. Usually the information of one is "linked" to the information of the other.

Metasearch engine

A search tool that uses more than one search engine.

Mosaic

A Web browser that was a precursor to Netscape Navigator.

Netscape Navigator

A Web browser. The main author is Mark Andreessen. He and Jim Clark founded Mosaic Communications, which became Netscape Communications Corporation.

Newsgroup

The name for discussion groups on Usenet.

Search engine

A tool that allows you to search the World Wide Web.

Server

A piece of software package or the machine it is operating on.

URL (Uniform Resource Locator)

An address of a Web site.

Usenet

A worldwide system used for discussion.

Web site

The page(s) contained at a specific URL.

WWW (World Wide Web)

A worldwide computer information system on the Internet.

INDEX

About the Author

JUDITH A. BAZLER is Associate Professor of Science Education in the Department of Curriculum and Instruction at Monmouth University.